THE FORMULA MANUAL

THE FORMULA MANUAL

Fourth Edition

by
STARK RESEARCH ASSOCIATES

ANDREWS AND McMEEL, INC.
A Universal Press Syndicate Company
KANSAS CITY • NEW YORK • WASHINGTON

CAUTION TO READERS

In writing *The Formula Manual,* I have excluded hazardous materials wherever possible, but in some cases they must be included to make a product effective. In making a formula, the reader should observe any note of caution added at the end of the recipe. In addition, the reader should consult Appendix 8 *(Sources of Supply and Index of Chemicals)* for the cautions specific to an individual ingredient.

We all know that materials such as waxes and oils will burn, so I have concentrated the warnings to materials that may be less familiar. But remember, *all* chemicals, including ordinary table salt, should be kept out of the reach of children, carefully labeled, and used only for the purpose they are intended.

The value and safety to you of the products in this book depend upon your careful use of the materials shown in the proportions given, as well as your observing any special cautions appearing in the book or with the materials. Neither I nor the publisher can be responsible for the efficacy of the products or your own safety if you do not follow these instructions and precautions.

CONTENTS

"All things are poisons . . . " said the sixteenth century scientist Paracelsus. "Only the dose decides that a thing is not poisonous."

Basically, all chemicals are potentially dangerous when improperly used. A good example that can be cited is sodium chloride, or common table salt. This chemical can be seriously dangerous when taken in excess. However, I doubt that few of us, unless restricted by medical diet, have eaten many meals without the use of the mineral. It is so essential in our physiological makeup that the human body cannot function long when there is an improper balance of salt in the system.

Hazardous materials have been excluded in this manual whenever possible, but in many cases they must be included to make a product effective. In preparing any of these formulas, the reader should always observe any note of caution added at the end of the recipe. Warnings pertaining to chemicals that may be dangerous are repeated in Appendix 8, *Sources of Supply and Index of Chemicals.* The reader should consult this appendix for the cautions pertinent to that ingredient before formulating any product to be sure of any hazardous properties.

To the best of our knowledge, no chemical has been recommended in these formulas which is banned by any of the various state or federal agencies. However, there is constant screening and monitoring of chemicals that are used in today's society and the approved list is constantly subject to change. If there is doubt concerning a particular compound, it is suggested that you contact your nearest Environmental Protection Agency or Federal Drug Administration office to confer with them on the subject.

We cannot overemphasize the importance of proper labeling of all formulas prepared with the date of preparation—the ingredients contained in them, and cautions pertaining to the ingredients. Labels have been provided for this purpose, so mix up the formula for Paste and use labels on all your finished products.

The value and safety to you or the user of your products from the formulas depend upon your careful use of the materials in the given proportions, as well as your observing the precautions noted throughout the manual. Neither I nor the publisher can be responsible for the efficacy of the products or your own safety. Because we have no control over the reader's procedures and methods of formulating, we must disclaim any responsibility for the use of this manual.

This new fourth edition of *The Formula Manual* is an extension of the first, second, and third editions that found wide acceptance among teachers and students throughout the United States and in foreign countries.

In the past three years, thousands of teachers both here and abroad, from elementary to college levels, have found *The Formula Manual* to be extremely helpful in courses ranging from elementary science to home economics, from basic chemistry to a continuing education course called "Consumer Chemistry," where it was used as the main text. By enabling students to make compounds that they are already familiar with around the home and ones that they can actually use, many basic concepts of chemistry are given immediacy, much of the mystery is removed from one level of their existence, and the way is opened to the study and contemplation of much more interesting and profound scientific mysteries if their inclination should lead there.

Aside from its use as a textbook, *The Formula Manual* has found its way into many kitchens and home workshops. There, people who are living on fixed incomes or are interested in making their own products as a hobby, have found that it has enabled them to save hundreds of dollars. We are, of course, delighted that the book has found such wide appeal and application.

The fourth edition has been expanded to include over 500 formulas. And, as in the third edition, the formulas are grouped into categories for ease in locating them. In addition to the new formulas, a number of existing ones have been modified as a result of improvements learned from the research and testing programs of past years. The appendices have been retained with additions and corrections being made where needed.

As an added help to those who teach the elementary grades, we have selected a group of formulas that we feel are appropriate for this age level; these are identified with the letter E following the formula in the index.

The Formula Manual is authoritative. Stark Research Corp. has maintained a full-scale research, development, and testing facility on the outskirts of Tucson, Arizona. Here new formulas have been developed and tested and existing ones have been updated and modified. The results of this effort have then been incorporated in later editions.

We have included many of the suggestions received from users of the first three editions and have worked hard to improve this one, but of course we realize there may be some imperfections in the formulas. Suggestions or criticisms that our readers might have for improving future editions are solicited.

A change in our method of operation has been made beginning with this fourth edition of *The Formula Manual*. The first, second, and third editions were published and sold by Stark Research Corp. Now, however, we find that the administrative duties inherent to doing our own publishing and selling could possibly detract from research time and the development of proposed collateral teaching materials. Accordingly, we arranged with Andrews & McMeel Publishers to take over the publishing and promoting of *The Formula Manual*.

The Formula Book is a condensed version of *The Formula Manual* and contains only a small portion of the formulas that are in this textbook. *The Formula Book* is intended to be a consumer edition—not a teaching aid. First published in November 1975, by Andrews & McMeel Inc., it became an immediate "best-seller." We asked them to publish *The Formula Manual*, as well, and they agreed.

Stark Research Corp.

The usage of abbreviations is so variable that no doubt our choice will not please everyone. We hope, at best, that our choices will be universally intelligible. They are:

teaspoon—t.	gallon—gal.
tablespoon—T.	pound—lb.
cup—C.	gram—g.
pint—pt.	kilogram—kg.
quart—qt.	milliliter—ml.
	liter—L.

Other less common measurements are spelled in full and all are discussed in either Appendix 6 or 7.

"I guess we should have tried it on the rats first."

BOOK COVER COATING

Ingredients:			
White Shellac	1 pt.	474 ml.
Denatured Alcohol	1 pt.	474 ml.

Mixing: Stir together white shellac and denatured alcohol.

Use: For protection against wear and mildew, apply to book covers with a fine brush. Be careful not to get any on the edges of the pages.

CAUTIONS: Denatured alcohol may be toxic if taken internally. Flammable.

Notes:

CANE CHAIR SEAT TIGHTENER

Ingredients:			
White Distilled Vinegar	3 C.	711 ml.
Water	1 C.	237 ml.
Powder Alum	2 T.	28 g.

Mixing: Mix vinegar and water by stirring and stir in powdered alum.

Use: Paint a liberal amount of this solution on cane chair seats and allow to dry. It is very important that the drying is done as quickly as possible to effect the greatest degree of shrinkage. So it is advisable that the chair be placed in direct sunlight if possible.

Notes:

CAST IRON CEMENT

Ingredients:			
Iron Filings	9 oz.	252 g.
Ammonium Chloride	1/2 oz.	14 g.
Powdered Sulfur	1/4 oz.	7 g.

Mixing: Dry mix iron filings, ammonium chloride, and powdered sulfur.

Use: Add water to form a paste the consistency you desire. Apply to cracks in cast iron and allow to harden.

CAUTIONS: Ammonium chloride is highly toxic by inhalation and ingestion and a strong irritant to tissue. If it comes in contact with skin, flush freely with clear water for prolonged periods of time.

Notes:

"These fireplace logs are just pieces of a tree. Haven't you got any of the real plastic ones?"

CEMENT WATERPROOFING

Ingredients: Ammonium Stearate   2-1/2 lb.   1.1 kg.
            Water   8 gal.   30.0 L.

Mixing: Mix ammonium stearate into water.

Use: Apply to any concrete or masonry surface with a large brush, such as a whitewash brush. Two or three coats are desirable.

Notes:

CHARCOAL STARTER

Ingredients: Mineral Oil   2-3/4 C.   651 ml.
            Kerosene   1/4 C.   59 ml.

Mixing: Mix mineral oil and kerosene together. Store in air tight containers.

Use: Pour over charcoal, saturating well, and ignite.

CAUTIONS: Kerosene is toxic and flammable. Use with caution.

Notes:

CHEMICAL FLOWER GARDEN

Ingredients: A Porous Brick
            A Shallow Pan or Transparent Bowl

Sodium Chloride (Salt)	12 T.	168 g.
Water	12 T.	180 ml.
Laundry Bluing	12 T.	180 ml.
Household Ammonia	2 T.	30 ml.
Red, Blue or Green Ink	8 Drops	1 ml.

Mixing: Mix the salt, water, laundry bluing and ammonium together. Add your choice of ink colors in spots on top of the brick.

Use: Place brick in the pan or bowl and fill 2-inches deep with the solution. Capillary action will cause the solution to migrate to the surface of brick where colored crystals will grow.

CAUTIONS: Avoid vapors of household ammonia.

Notes:

DISINFECTANT FOR GENERAL USE

Ingredients:			
Cresylic Acid	1-1/4 C.	296 ml.
Oleic Acid	3 T.	44 ml.
Caustic Soda	1 t.	4 g.
Water	1/2 C.	118 ml.
Sulphonated Castor Oil	2/3 C.	158 ml.

Mixing: Warm cresylic acid but do not boil, and stir in oleic acid. Separately, dissolve the caustic soda into water. Mix the two solutions together and add sulphonated castor oil, mixing thoroughly.

Use: Use either full strength or diluted with water to the strength you desire. This can be used in kitchen, bathroom, animal living quarters, or wherever desired.

CAUTIONS: Caustic soda heats on contact with water and can cause severe burns to skin. Handle with care. Store in airtight container. Cresylic acid is toxic and can be absorbed through skin; use gloves in formulating disinfectant.

Notes:

DISINFECTANT FOR SHOES

Ingredients:			
Formaldehyde	1 T.	15 ml.
Water	1/4 C.	59 ml.

Mixing: Simply stir formaldehyde into water.

Use: With a cotton ball or piece of absorbent cloth, swab out inside of shoes three times, allowing to dry between each application.

CAUTIONS: Formaldehyde is highly toxic by inhalation, ingestion, and skin contact. Handle with care, do not breathe fumes, and if splashed accidentally on skin, flush with plenty of clear water.

Notes:

DRAIN CLEANER

Ingredients:			
Sodium Bicarbonate	1 C.	227 g.
Sodium Chloride	1 C.	227 g.
Cream of Tartar	1/4 C.	57 g.

Mixing: Blend the three powders together with a wooden fork or spoon. Store in an airtight container.

Use: Pour 1/4 cup (57 g.) of mixture in drain and add 1 cup (237 ml.) boiling water. After a minute, flush with cold water. This will not open a clogged drain, but weekly use will keep drains free. Note the bubbling action caused by a base neutralizing an acid.

Notes:

DRAIN OPENER

Ingredients:		
Caustic Soda	3/4 C.	170 g.
Calcium Carbonate	1 C.	227 g.
Caustic Potash	3 C.	681 g.

Mixing: Blend ingredients together with wooden fork or spoon, and store in an airtight container.

Use: Put 2 tablespoons (28 g.) of mixture in drain, followed by 1 cup (237 ml.) boiling water. Let stand for 30 minutes, then flush with cold water.

CAUTIONS: This is a **highly caustic mixture**, do not allow contact with skin. If it does get on skin, or some of the drain water splashes on hands or arms, flush with plenty of clear water. Caustic soda and caustic potash both heat on contact with water and can cause severe burns to skin. Handle with care, wear rubber gloves. Store in airtight container.

Notes:

FABRIC FLAMEPROOFING COMPOUND

Ingredients:		
Boric Acid	3/4 lb.	340 g.
Borax	1 lb.	454 g.
Water	2 gal.	8 L.

Mixing: Mix boric acid, borax, and water.

Use: Immerse fabric in mixture and thoroughly saturate. Spin dry and complete drying. This is a water-soluble solution so it must be treated every time the material is washed.

Notes:

FIBERGLASS FILTER TREATMENT

Ingredients:			
Paraffin Oil	5 C.	1.2 L.
Stearic Acid	2 C.	454.0 g.
Bentonite	1-1/2 C.	340.0 g.
Triethanolamine	3/4 C.	177.0 ml.
Water	2-1/2 gal.	10.0 L.

Mixing: Mix paraffin oil, stearic acid, and bentonite, stirring well. Separately mix triethanolamine and water, then combine the two mixtures, stirring until a cloudy emulsion forms.

Use: If you have a plain, untreated fiberglass filter, spray this on it before using. This mixture greatly improves its dust-catching ability. If not used immediately, stir well before using.

CAUTIONS: Triethanolamine may be somewhat irritating to skin and mucous membranes.

Notes:

FILTER CLEANING COMPOUND

Ingredients:			
Sodium Metasilicate	1 t.	4 g.
Detergent, liquid	1 t.	5 ml.
Water	1 qt.	1 L.

Mixing: Mix sodium metasilicate and liquid detergent into water.

Use: Brush as needed onto wet filter and hose off with a strong stream of water. Dry and reinstall.

Notes:

FIRE EXTINGUISHING LIQUID

Ingredients:			
	Sodium Carbonate	2 C.	454.0 g.
	Alum	1 C.	227.0 g.
	Borax	3/4 C.	170.0 g.
	Potassium Carbonate (Potash)	1/4 C.	57.0 g.
	Sodium Silicate	3 pts.	1.4 L.
	Water	1 gal.	3.8 L.

Mixing: Mix sodium carbonate, alum, borax, and potassium carbonate into the sodium silicate. Use 3 cups (681 g.) of the resulting mixture in 1 gallon (3.8 L.) of water.

Use: Use in a coarse spray or sprinkler can, directing the spray at the base of the fire first to contain it. Then work gradually until the fire is thoroughly extinguished and all sparks are out.

CAUTIONS: Potassium carbonate (potash) is toxic if taken internally. Sodium silicate may be irritating and caustic to skin and mucous membranes.

Notes:

FIRE EXTINGUISHING POWDER

Ingredients:			
	Fine Silica Sand (Mason Sand)	6 lb.	2.7 kg.
	Sodium Bicarbonate	2 lb.	900.0 g.

Mixing: Stir silica sand and sodium bicarbonate together.

Use: Sprinkle directly on small fires as needed.

Notes:

FIREPLACE FLUE SOOT REMOVER

Ingredients:			
	Sodium Chloride (Salt)	1 C.	227 g.
	Zinc Oxide Powder	1 C.	227 g.

Mixing: Mix together with a wooden fork or spoon the sodium chloride and zinc oxide and store in an airtight container.

Use: Sprinkle 1 cup (227 g.) on a hot fire, repeat in a few minutes.

CAUTIONS: Zinc oxide is highly poisonous if taken internally.

Notes:

FIREPLACE STARTER

Ingredients:			
Sawdust		5 lb.	2.25 kg.
Fuel Oil or Auto Crankcase Oil		1 qt.	1.00 L.

Mixing: Mix the oil into the sawdust slowly and thoroughly, allowing the sawdust to completely absorb the oil. Store in an airtight can.

Use: Sprinkle a small amount on and around fireplace logs and ignite.

CAUTIONS: Fuel oil is flammable.

Notes:

FIREPROOFING FOR PAPER

Ingredients:		
Ammonium Sulfate	1 C.	227 g.
Boric Acid	6 T.	84 g.
Borax	4 T.	56 g.
Water	3 C.	711 ml.

Mixing: Dissolve ingredients in water, mixing well.

Use: Paper to be protected may be dipped in solution, or it may be brushed on specific areas to be protected. Use several coats, allowing each one to dry thoroughly before applying the next.

Notes:

FIREPROOFING SYNTHETIC FABRICS

Ingredients:		
Boric Acid	1 C.	227.0 g.
Water	1 gal.	3.8 L.

Mixing: Dissolve boric acid in water.

Use: Immerse fabric in mixture, allow to soak, wring out, and hang up to dry. Retreat fabric after each laundering. This may be done by adding boric acid to the final rinse cycle of the washing machine.

Notes:

FIREPROOFING TEXTILES

Ingredients: Ammonium Phosphate   1/2 C.   113.0 g.
Ammonium Chloride   1 C.   227.0 g.
Water   3 pt.   1.4 L.

Mixing: Stir ammonium phosphate and ammonium chloride into water.

Use: Soak cloth (tents, banners, or other textiles) in solution for a few minutes, wring out, and hang up to dry. Cloth must be retreated after each exposure to water.

Notes:

FIREPROOFING THE CHRISTMAS TREE

Ingredients: Ammonium Sulfate   1 C.   227.0 g.
Boric Acid   1/2 C.   113.0 g.
Borax   2 T.   28.0 g.
Water   1 gal.   3.8 L.

Mixing: Mix ammonium sulfate, boric acid, and borax into water.

Use: Spray the tree with this solution, and use it also to fill the cup of the tree stand.

Notes:

FIREPROOFING FOR WOOD

Ingredients: Zinc Chloride   1/2 C.   113 g.
Ferric Chloride   1/4 C.   57 g.
Boric Acid   3 T.   42 g.
Ammonium Phosphate   3 T.   42 g.
Water   2 qt.   2 L.

Mixing: Stir chemicals into water, mixing well.

Use: Spray or paint on areas to be protected. Use three or four coats for best results.

CAUTIONS: Ferric chloride is slightly toxic.

Notes:

FIREWOOD SUBSTITUTE

Ingredients: Old Newspapers
Library Paste
Broomstick—about 36" long
Masking Tape or Rubber Bands

Making: Separate a stack of newspapers into single sheets, leaving them folded in the middle to the regular size. Coat the broomstick with silicone emulsion, or a heavy coat of wax or vegetable oil. Dab spots of paste on the top of the first sheet of newsprint and roll it around the stick. Using just enough paste to hold it all together, continue in the same way until the roll is about 4 inches in diameter. Then secure the roll with tape or rubber bands until the paste is dry. When dry, remove the stick from the center and you have a newspaper log that will burn quite well in the fireplace. If the paste is too thick, thin it with water so that it will brush on easily. (The formula for Library Paste can be found later in this section.)

Use: Burn as logs in the fireplace.

Notes:

FUEL OIL IMPROVER

Ingredients:			
Naphthalene	3 qt.	2.8 L.
Anthracene	1-1/4 C.	296.0 ml.

Mixing: Mix the naphthalene into the anthracene, stirring well.

Use: Add to fuel at the rate of 2-1/2 cups (592 ml.) per 100 gallons (380 L.). This will help prevent soot and gum deposits on the burners, keeping them clean and efficient.

CAUTIONS: Anthracene is hazardous—it is toxic and an irritant. Naphthalene is moderately toxic.

Notes:

IRON RUSTPROOFING COMPOUND

Ingredients:			
	Sodium Silicate	40 oz.	1200 ml.
	Trisodium Phosphate	3/4 oz.	21 g.
	Soda Ash	1/4 oz.	7 g.
	Water	2 qt.	2 L.

Mixing: Stir sodium silicate, trisodium phosphate, and soda ash into water.

Use: With mixture at room temperature, liberally coat a piece of iron and wipe dry.

CAUTIONS: Trisodium phosphate is moderately toxic by ingestion. Both trisodium phosphate and sodium silicate may be irritating and caustic to skin and mucous membranes. Rubber gloves should be worn when handling these.

Notes:

IRON RUST-STAIN REMOVER FOR CONCRETE

Ingredients:			
	Sodium Citrate	1/2 C.	113 g.
	Glycerin	3 C.	711 ml.
	Water	3 C.	711 ml.
	Chalk (Calcium Carbonate)	(See mixing instructions)	

Mixing: Mix sodium citrate and glycerin into water. Stir in enough chalk until a doughlike consistency results.

Use: Trowel dough onto stained area and allow to stand from one to three days. Remove and scrape off material.

Use: For stained concrete.

CAUTIONS: Sodium citrate is combustible.

Notes:

LAMP AND TORCH OIL

(This formula is for deodorized kerosene that will burn cleanly.)

Ingredients:		
Kerosene	1 gal.	3.8 L.
Mineral Lime	3 oz.	84.0 g.
Oil base Dye (optional)	to suit	

Mixing: Mix mineral lime into kerosene, stirring thoroughly. Strain into clean container through several layers of cheesecloth, and add an oil base dye, if desired, for lamps with glass reservoirs.

Use: Use in indoor kerosene lamps or outdoor patio torches.

CAUTIONS: Kerosene is toxic if taken internally and flammable. Wear rubber gloves while handling.

Notes:

LIQUID LIBRARY ADHESIVE

Ingredients:		
Yellow Dextrin	4 T.	56 g.
Calcium Chloride	2 T.	28 g.
Water	2 C.	474 ml.

Mixing: Mix ingredients together at room temperature. Heat slowly while stirring until a thin, uniform consistency has been reached. Be careful not to boil. Thickness can be adjusted by varying the amount of water.

Use: Use for any cutting and pasting job.

Notes:

LIBRARY PASTE

Ingredients:		
Cornstarch	1/2 C.	113.0 g.
Cold Water	3/4 C.	177.0 ml.
Boiling Water	6 C.	1.4 L.

Mixing: Mix cornstarch and cold water to form a paste. Add to boiling water and stir until a translucent mixture forms.

Use: Use as above for a less permanent bond.

Notes:

LUBRICATING STICK

Ingredients: White Petrolatum   1/2 C.   113 g.
             Paraffin Wax   1/2 C.   113 g.

Mixing: Put ingredients in the top of a double boiler and heat until melted. Stir together well. Roll a piece of heavy paper over a round peg and secure it with tape to form a mold. Fold one end over and secure with tape to close. Remove peg. Pour mixture into molds when cool, but still pourable.

Use: Rub on drawer frames, window sashes, or any sliding surface.

Notes:

LUMINARIAS

(This is a remarkably beautiful outdoor lighting effect that originated in Mexico and is now used widely throughout the United States, especially at Christmas.)

Ingredients: Brown Paper Bags, medium size—as many as you want
             Sand—enough to have 3 or 4 inches in the bottom of each bag
             Candles—one for each bag

Making: With a scoop, put 3 or 4 inches of sand in the bottom of each bag, push the candles into the sand so that they stand upright in the center. Light the candles, leaving the top of the bags open. If you fold the sides of the bags over in a little cuff they will hold their shape better.

Use: Use them for outdoor evening parties. Line driveways and walks with them.

Notes:

MOTHPROOFING FOR TEXTILES

Ingredients: Ammonium Selanate   1 t.   4.0 g.
             Water   1 gal.   3.8 L.

Mixing: Mix ammonium selanate into water.

Use: Soak any washable material to be protected, wring out, and air dry.

CAUTIONS: Ammonium selanate may be mildly toxic.

Notes:

MOTH REPELLENT

Ingredients: Paradichlorobenzene   2 C.   454 g.

Mixing: Melt in the top of a double boiler and hold temperature between 130–133°F. (54–57°C.).

Use: Cut cardboard strips about 2 inches by 4 inches and dip repeatedly in the melted solution, allowing to cool between each dip, until a heavy coating is built up. Hang strips in closets or put them in clothing-storage areas.

CAUTIONS: Paradichlorobenzene is moderately toxic by ingestion and an irritant to eyes. Mix in well ventilated area.

Notes:

PINE OIL DISINFECTANT

Ingredients:		
Raw Pine Oil	2 C.	474 ml.
Sulfonated Castor Oil	1 C.	237 ml.
Oleic Acid	6 T.	89 ml.
Caustic Potash	1 t.	4 g.

Mixing: Mix together raw pine oil, sulfonated castor oil, and oleic acid. Then stir in caustic potash.

Use: Use about 1/2 cup (118 ml.) per gallon (3.8 L.) of warm water for mopping down floors, cleaning bathrooms, and so on.

CAUTIONS: Caustic potash heats on contact with water and can cause severe skin burns, so handle with care. If it splashes on skin, rinse with water. Wear rubber gloves.

Notes:

PLASTIC AIR FILTER FOR FURNACE AND AIR CONDITIONER

Ingredients: 1. Polyurethane Foam—cut to the size of your filter holder, not more than 1/2-inch in thickness, and less than 1 pound (454 g.) per cubic foot density.

2. Acetone   1 qt.   1 L.

Making: Pour the acetone in a shallow pan and immerse the foam sheet. It will swell quickly to 5 or 6 times the original thickness. This expansion ruptures many of the membranes between air cells in the plastic foam, so that air passes through it easily. Its thickness, and the small size of the tunnels and passages that the air must move through, make this an excellent filter.

Use: This filter may be used indefinitely. Be sure to wash it once a month under normal conditions to remove the collection of dust particles.

CAUTIONS: Acetone is extremely volatile and flammable. Use with caution away from your flame. Do not rub plastic together—a static charge may build up. Dip and squeeze gently. Allow the acetone to evaporate before installing the filter in the furnace.

Notes:

REMOVING DRIED PLASTER FROM HANDS

Ingredients: Sugar   1/2 C.   113 g.
            Water   1 C.   237 ml.

Mixing: Dissolve sugar into water, stirring to form a slurry.

Use: Apply to hands and rub in well. Flush with clear water.

Notes:

ROOT DESTROYER FOR DRAINS

Ingredients: Caustic Soda   3 C.   681 g.
            Copper Sulfate   1/4 C.   57 g.
            Ammonium Sulfate   1 t.   4 g.

Mixing: Mix the three ingredients with a spoon or wooden fork.

Use: Put 1 cup (227 g.) in drain, or down toilet. Flush with hot water if possible.

CAUTIONS: Caustic soda heats on contact with water and can cause severe skin burns. Handle with care. Store in airtight container. Copper sulfate is highly toxic. Wear rubber gloves.

Notes:

SEPTIC TANK REACTIVATOR

Ingredients:			
Brown Sugar	1 lb.	454 g.
Dried Yeast	1 envelope	
Water	1 qt.	1 L.

Mixing: Simply add the brown sugar and yeast to water, and stir until most of the sugar is dissolved.

Use: Pour mixture in the toilet and flush. This will restart the proper bacterial action in the septic tank.

Notes:

SNOW AND ICE-MELTING COMPOUND

Ingredients:		
Rock Salt	4 C.	908 g.
Ammonium Sulfate	1/8 C.	32 g.
Magnesium Sulfate	4 C.	908 g.

Mixing: Mix the three ingredients together thoroughly and store in an airtight container.

Use: Sprinkle about one handful per six-square feet on snow or ice.

Notes:

SOAP SALVAGE

A significant portion of each bar of soap is wasted because it is too small for easy handling. Here is a simple, economical, conservation method for utilizing normally wasted soap ends.

Cut the ends you've saved into slivers and put them in a pan with a small amount of water, just enough to cover. Heat to a boil and reduce by roughly the amount of water you put in. Cool down and pour into molds. Plastic soap dishes are good, reusable molds. Allow to set for several days, remove from molds, and use.

Notes:

TOBACCO PIPE CLEANER AND DEODORANT

Ingredients:			
	Isopropyl Alcohol	3 T.	45 ml.
	Water	10 T.	150 ml.
	Soap Chips	2 T.	28 g.
	Formaldehyde	2 T.	30 ml.

Mixing: Mix the isopropyl alcohol with the water and add soap chips to the mixture. Then add formaldehyde as the final step and stir.

Use: Remove built-up residue in bowl of pipe to desired level, then swab pipe and bowl with a pipe cleaner saturated in this solution.

CAUTIONS: Formaldehyde is highly toxic by ingestion, inhalation, or skin contact. Isopropyl alcohol is mildly toxic by inhalation and ingestion. Flammable.

Notes:

UTILITY CANDLES

In light of the energy problem and the resulting brownouts and blackouts, the utility candle, like the oil or kerosene lamp, has become a necessity in the home. Here's a formula for a good serviceable and inexpensive candle that is easily made.

Ingredients:			
	Paraffin Wax	2 C.	454 g.
	Stearic Acid	1 C.	227 g.
	Cotton Wicking	several feet	
	Molds	several	

Making: The simplest mold for a square candle is a quart-size milk carton with the top cut off. For round candles, use a section of paperboard mailing tubes, or the tubes from the inside of toilet paper rolls or paper towels, with one end sealed by taping a paperboard disc in place.

Cut a length of wick and tie one end around a pencil or small stick that will go across the top end of the mold. Tie a metal nut or washer to the other end so that it hangs just at the bottom of the mold. This will keep the wick straight and centered.

Melt paraffin wax and stearic acid in the top of a double boiler, allow to cool to just above solidification point, and pour into molds. At first, pour just a little in the bottom to make sure there are no leaks. As the wax cools, it will contract a little, leaving a depression in the center around the wick. If necessary, fill it up with more melted wax. When the wax has hardened, cut away the paper mold. If a smooth, shiny surface is desired, hold the candle by the wick and dip it in and out of boiling water and then immediately dip it in ice water.

If desired, the wick can be put in after the candle has been melted by making a hole down the center of the candle with a small, long drill, or a heated rod—such as an ice pick.

Be sure not to use force, as this will break the candle. When the hole is made, dip the wick in melted wax and cool, stretched out between two pins. When it is cool, gently push the wick into the candle. This method might be one way to find the right size wick to use for your candles—if the wick is too thick, it will burn too fast and wax will drip down the sides. If it is too thin, it won't burn up the wax fast enough and the wick will drown. Generally, the larger the diameter of the candle, the thicker the wick should be.

Notes:

WALLPAPER REMOVING

Ingredients:			
Liquid Detergent	1/2 C.	118 ml.
Water	2 C.	474 ml.
Ethylene Glycol Monoethylether	1/2 C.	118 ml.

Mixing: Mix the liquid detergent with water, then gently stir in the ethylene glycol monoethylether.

Use: Mix 1/2 cup (118 ml.) to 1 quart (1 L.) warm water. Apply with sponge and allow to soak in for 10 minutes before taking wallpaper off with scraper.

CAUTIONS: Ethylene glycol monoethylether is flammable. Mix ingredients in a well ventilated room.

Notes:

"Look at my wall! Can't you tell the difference
between bug-killer and spray-paint?"

ANTISTATIC SPRAY FOR RUGS

Ingredients:			
Silicone Oil Emulsion	3 T.	45 ml.
Water	1 qt.	1 L.

Mixing: Mix silicone oil emulsion into water.

Use: Spray on rug or carpet in heavy traffic areas to prevent static electricity shocks. Each treatment will be effective for three or four weeks, depending on traffic and the type of fabric. This compound works by reducing the friction between your shoes and the rug.

Notes:

CARPET CLEANER

Ingredients:		
Whole Wheat Flour	3 C.	681 g.
Water	1-3/4 C.	414 ml.
Aluminum Stearate	1 T.	14 g.
Salicylic Acid	1 T.	14 g.
Mineral Oil	1-1/4 C.	296 ml.

Mixing: Mix whole wheat flour and water into paste. Add aluminum stearate and salicylic acid, mixing well. Then add the mineral oil and mix well.

Use: Brush paste into soiled area and wipe away with a damp cloth and clear water.

Notes:

CEMENT FLOOR HARDENER AND SEALER

Ingredients: Magnesium Fluosilicate	1 lb.	454 g.
Water	2 gal.	8 L.

Mixing: Mix the magnesium fluosilicate with water.

Use: Flush over surface of floor with a broom and allow to stand for one hour. Wash off with clear water.

CAUTIONS: Magnesium fluosilicate is highly toxic when taken internally and is a strong irritant to skin. Use only in well ventilated area and wear rubber gloves.

Notes:

CERAMIC TILE CLEANER

Ingredients: Trisodium Phosphate	2 T.	28.0 g.
Water	1 gal.	3.8 L.

Mixing: Mix trisodium phosphate into water.

Use: Apply to tile floor with brush, sponge, or cloth. Wring out a sponge and mop-up dirty water. The floor needs no rinsing, unless higher concentrations of trisodium phosphate are used, for heavy-duty cleaning.

CAUTIONS: Trisodium phosphate is a skin irritant; use rubber gloves. It is also moderately toxic by ingestion.

Notes:

CONCRETE ACID WASH

Ingredients: Aluminum Chloride	1.5 lb.	681 g.
Water	10 gal.	38 L.

Mixing: Mix aluminum chloride into water.

Use: Apply to concrete surface with brush or broom. After one hour, wash with clear water.

CAUTIONS: Aluminum chloride is highly toxic by inhalation and ingestion and a strong irritant to tissue. If it comes in contact with skin, wash with clear water for prolonged periods of time. Wear rubber gloves.

Notes:

CONCRETE CLEANER

Ingredients:		
Sodium Metasilicate	3-1/4 C.	738 g.
Trisodium Phosphate	3/4 C.	170 g.
Soda Ash	1/2 C.	113 g.

Mixing: Stir all the ingredients together, mixing well.

Use: Wet concrete surface, sprinkle mixture around, allow to stand for 15 minutes, then rinse with clear water.

CAUTIONS: Trisodium phosphate is a skin irritant; use rubber gloves. It is also moderately toxic by ingestion.

Notes:

CONCRETE FLOOR DUSTPROOFING

Ingredients:		
Sodium Silicate (Waterglass)	1 qt.	1.0 L.
Water	3 qt.	2.8 L.

Mixing: Mix sodium silicate into water.

Use: Mop the concrete floor with plain water. When just dry or still slightly damp apply a liberal coat of the mixture and allow 24 hours before using.

CAUTIONS: Sodium silicate may be irritating and caustic to skin and mucous membranes.

Notes:

©1976 Universal Press Syndicate   12/29

"You're washing the floor with tomorrow's soup-of-the-day!"

LIQUID FLOOR WAX EMULSION

Ingredients:			
Carnauba Wax		30 oz.	850 g.
Castile Soap		7-1/2 oz.	210 g.
Oleic Acid		1-1/2 oz.	45 ml.
Caustic Soda		3/4 oz.	21 g.
Water		1.5 gal.	6 L.

Mixing: Melt carnauba wax, castile soap, and oleic acid in top of double boiler. In a separate container, dissolve caustic soda and water. When the ingredients in the double boiler have melted, turn off heat and add caustic soda and water mixture in a steady stream with rapid stirring, continuing until the liquid becomes milky—this indicates that the emulsion has formed.

Use: Apply to floor and spread with applicator. When floor is dry, buff it.

CAUTIONS: Caustic soda heats on contact with water, is highly caustic, and can cause skin burns. In the event of contact with skin, flush freely with clear water for prolonged periods of time. Handle with care and store in airtight container.

Notes:

DANCE FLOOR WAX

Ingredients:			
Stearic Acid		1/2 lb.	227 g.
Talc		1/2 lb.	227 g.

Mixing: Mix together stearic acid and talc, stirring well.

Use: Apply lightly with a shaker to floor.

Notes:

WOOD FLOOR BLEACH I

Ingredients:			
Sodium Metasilicate		9 C.	2 kg.
Sodium Perborate		1 C.	227 g.

Mixing: Stir ingredients together thoroughly.

Use: Use 1 pound (254 g.) of mixture to 1 gallon (3.8 L.) boiling water. Spread on floor with a mop, allow to stand for 30 minutes, and rinse with clear water. Use rubber gloves.

Notes:

WOOD FLOOR BLEACH II

Ingredients: Sodium Perborate   1/2 lb.   227 g.
              Water   enough to make paste

Mixing: Mix enough water into sodium perborate to form a thin paste.

Use: Rub on floor, let stand for 30 minutes, and rinse. Use rubber gloves.

Notes:

WOOD FLOOR CLEANER

Ingredients: Mineral Oil   2-1/4 C.   533 ml.
              Oleic Acid   3/4 C.   177 ml.
              Ammonia   2 T.   30 ml.
              Turpentine   5 T.   75 ml.

Mixing: Mix mineral oil and oleic acid together, then add ammonia and turpentine.

Use: Mix 1 cup (237 ml.) to 8 cups (1.9 L.) water. This is excellent for cleaning parquet floors with a sponge mop or coarse towelling.

CAUTIONS: Turpentine is toxic if taken internally. It is also flammable. Handle with care. Avoid vapors of ammonia.

Notes:

FLOOR POLISH

Ingredients: Paraffin Wax   1/4 C.   57 g.
              Mineral Oil   2 qt.   2 L.

Mixing: Melt paraffin wax in the top of a double boiler; stir in mineral oil. Cool and store in bottles.

Use: Just dampen a cloth or sponge with polish and go over floor.

Notes:

LIQUID FLOOR WAX I

Ingredients:			
Yellow Beeswax	2 T.	28 g.
Ceresin Wax	1/2 C.	113 g.
Turpentine	2 C.	474 ml.
Pine Oil	1 T.	15 ml.

Mixing: Melt yellow beeswax and ceresin wax in double boiler. Turn off heat and cool for a few minutes, then slowly stir in turpentine and pine oil. Let cool to room temperature and pour into containers.

Use: Apply a thin coat to floors, let dry, and polish with a clean dry cloth or an electric buffer.

CAUTIONS: Turpentine is toxic if taken internally. It is also flammable. Handle with care.

Notes:

LIQUID FLOOR WAX II

Ingredients:			
Beeswax	1/4 lb.	113 g.
Paraffin Wax	1 lb.	454 g.
Raw Linseed Oil	1/2 C.	118 ml.
Turpentine	3 C.	711 ml.

Mixing: Melt beeswax and paraffin wax in top of double boiler. Turn off heat and let cool for a few minutes. Stir in linseed oil and turpentine, stirring rapidly until the whole is well mixed.

Use: Apply a thin coat to floors, let dry, and polish with a clean cloth or an electric buffer.

CAUTIONS: Turpentine is toxic if taken internally. It is also flammable. Handle with care.

Notes: Linseed oil dries when exposed to air. Keep in airtight container.

FLOOR-SWEEPING COMPOUND

Ingredients:			
	Sawdust (sifted)	6 C.	1.4 kg.
	Rock Salt	2 C.	454.0 g.
	Mineral Oil	1 C.	237.0 ml.

Mixing: Mix sawdust with rock salt, and then mix in mineral oil. Stir well to allow complete absorption of the mineral oil.

Use: Sprinkle on the floor, then sweep. This keeps down dust as you work.

Notes:

LINOLEUM PASTE WAX

Ingredients:			
	Carnauba Wax	1-1/2 C.	340 g.
	Ceresin Wax	1-1/2 C.	340 g.
	Mineral Spirits	4 C.	948 ml.

Mixing: Melt carnauba and ceresin waxes in a double boiler. Turn off heat and allow to cool for a few minutes, then slowly stir in mineral spirits. When the mixture just begins to solidify around the edges, pour into a container.

Use: Apply with a cloth or sponge, let dry, and polish with a soft cloth or buffer.

CAUTIONS: Mineral spirits is flammable.

Notes:

LINOLEUM POLISH

Ingredients:			
	Carnauba Wax	1/2 C.	113 g.
	Paraffin Wax	2 T.	28 g.
	Yellow Beeswax	4 T.	56 g.
	Turpentine	4 C.	948 ml.

Mixing: Melt the three waxes in a double boiler. Turn off heat and allow to cool for a few minutes. Then stir in turpentine and cool to room temperature. Store in bottles.

Use: Apply a light coat with a cloth or sponge, let dry, and polish with a soft cloth or buffer.

CAUTIONS: Turpentine is toxic if taken internally. It is also flammable. Handle with care.

Notes:

MINERAL OIL EMULSION I

Ingredients:			
Mineral Oil		4 C.	948 ml.
Ammonium Oleate		6 T.	84 g.
Water		3 C.	711 ml.

Mixing: Warm all ingredients in a double boiler, then mix vigorously with an eggbeater or electric mixer until a milky emulsion forms.

Use: Apply with a sponge or cloth to tile floors. This is excellent for use on Mexican or quarry tiles, or brick.

Notes:

MINERAL OIL EMULSION II

Ingredients:			
Mineral Oil		2-1/2 C.	592 ml.
Oleic Acid		3 T.	45 ml.
Triethanolamine		2 T.	30 ml.
Water		3 C.	711 ml.

Mixing: Warm all ingredients in a double boiler, then mix vigorously with an eggbeater or electric mixer until a milky emulsion forms.

Use: Apply with a sponge or cloth to tile floors.

CAUTIONS: Triethanolamine may be somewhat irritating to skin and mucous membranes.

Notes:

MOP OIL

Ingredients:			
Turpentine		1 C.	237 ml.
Mineral Oil		2 C.	474 ml.

Mixing: Stir the turpentine into the mineral oil.

Use: Barely dampen a dustmop with a few drops of this oil so that it will collect dust but not leave oil on the floor.

CAUTIONS: Turpentine is toxic if taken internally and flammable. Handle with care.

Notes:

ALUMINUM CLEANER

Ingredients: Aluminum Potassium Sulfate

(Powdered Alum)	2 T	28 g.
Trisodium Phosphate	1 C.	227 g.
Water	to form paste	

Mixing: Mix powdered alum and trisodium phosphate. Then add water, stirring slowly, until a thick paste is formed.

Use: Apply a small amount with a soft cloth to any aluminum product and rub lightly. Rinse off with clean water.

CAUTIONS: Trisodium phosphate is a skin irritant and moderately toxic by ingestion. Use rubber gloves.

Notes:

ALUMINUM POLISH I

Ingredients: Aluminum Potassium Sulfate

(Powdered Alum)	1 C.	227 g.
Calcium Carbonate (Chalk)	1-1/3 C.	303 g.
Talc	1 C.	227 g;

Mixing: Mix all three together well.

Use: Use a small amount on a damp, soft cloth, rub well, and rinse off with warm, clean water.

Notes:

ALUMINUM POLISH II

Ingredients:		
Paraffin Wax (Powdered)	1 oz.	28 g.
Magnesium Oxide	4 oz.	112 g.
Calcium Carbonate (Chalk)	3 oz.	84 g.
Iron Oxide (Red)	2 oz.	56 g.

Mixing: Mix all four ingredients together in a bowl, using a fork.

Use: Transfer to an open-mouth jar. Apply to aluminum with a damp cloth and rub. Wash with clear water.

Notes:

"It says right here on the can, 'Do not use
to clean overnight.'"

HOUSEHOLD AMMONIA

Ingredients: Ammonium Hydroxide   3/4 C.   177.0 ml.
             Water   1 gal.   3.8 L.

Mixing: Add ammonium hydroxide to water, with gentle stirring.

Use: Use as a general cleaning agent diluted with water, or in any formula where household ammonia is called for.

CAUTIONS: Vapors of ammonium hydroxide are irritating and corrosive to eyes and skin. It is toxic by ingestion. Handle carefully. If it is splashed on skin, rinse well with plenty of water. Also rinse utensils thoroughly after use. Wear rubber gloves.

Notes:

AMMONIA SUBSTITUTE

Ingredients: Trisodium Phosphate   6 T.   84.0 g.
             Lauryl Pyridinium Chloride   1/2 T.   7.0 g.
             Water   1 gal.   3.8 L.

Mixing: Stir trisodium and lauryl pyridinium chloride with water.

Use: Use this in the same way you would use regular ammonia. It is not as strong, however.

CAUTIONS: Trisodium phosphate is a skin irritant and moderately toxic by ingestion. Use rubber gloves. Lauryl pyridinium chloride may be mildly irritating to skin.

Notes:

AMMONIA WASHING POWDER

Ingredients: Powdered Soap   2 C.   454 g.
             Ammonium Carbonate   2 C.   454 g.

Mixing: Simply mix the two ingredients together.

Use: Add approximately 2 tablespoons (28 g.) to each quart of water for general household cleaning jobs.

CAUTIONS: Ammonium carbonate, if heated, may result in irritating fumes.

Notes:

ANTISEPTIC CLEANER, GENERAL USE

Ingredients:		
Trisodium Phosphate	6 C.	1.4 kg.
Soda Ash	1 C.	227.0 g.
Sodium Perborate	3/4 C.	170.0 g.
Borax	1/4 C.	57.0 g.
Powdered Soap	2 C.	454.0 g.

Mixing: Mix all five ingredients above together.

Use: Use 1 cup (227 g.) to 1 quart (1 L.) hot water for general cleaning.

CAUTIONS: Trisodium phosphate is a skin irritant and moderately toxic by ingestion. Use rubber gloves.

Notes:

ANTISEPTIC DETERGENT

Ingredients:		
Trisodium Phosphate	2 C.	454 g.
Sodium Bicarbonate (Baking Soda)	1 C.	227 g.
Sodium Pyrophosphate	1 C.	227 g.

Mixing: Mix all three together well.

Use: Use as a regular washing detergent whenever antiseptic action is required.

CAUTIONS: Trisodium phosphate is a skin irritant and moderately toxic by ingestion. Use rubber gloves.

Notes:

BATHTUB AND SINK CLEANER

Ingredients:		
Trisodium Phosphate	1 C.	227 g.
Powdered Soap	1 C.	227 g.
Calcium Carbonate (Chalk)	2 C.	454 g.

Mixing: Mix three ingredients together in a large bowl with a wooden fork or spoon.

Use: Pick up some of this powder on a damp cloth and rub on the area to be cleaned. Rinse with clear water.

CAUTIONS: Trisodium phosphate is a skin irritant and moderately toxic by ingestion. Use rubber gloves.

Notes:

BLACKBOARD CLEANER

Ingredients:			
Acetic Acid (Vinegar)		2 C.	474.0 ml.
Detergent		1 C.	227.0 g.
Water		1 gal.	3.8 L.

Mixing: Mix acetic acid and detergent into water.

Use: Wash blackboard with a cloth or sponge. Clean with a squeegee or rinse with clean water.

Notes:

BRASS AND COPPER CLEANER

Ingredients:			
Sodium Bicarbonate (Baking Soda)		1/4 C.	57 g.
Sodium Metasilicate		1/2 C.	113 g.
Trisodium Phosphate		1/4 C.	57 g.

Mixing: Mix all three ingredients together well.

Use: Use a small amount on a damp, soft cloth, rub well, and rinse off with warm, clean water.

CAUTIONS: Trisodium phosphate is a skin irritant and moderately toxic by ingestion. Use rubber gloves.

Notes:

BRASS PASTE POLISH

Ingredients:			
Stearic Acid or Powdered Paraffin		3/4 T.	10 g
Petroleum Distillate		1/4 C.	59 ml.
Caustic Soda		1/2 T.	7 g.
Denatured Alcohol or			
Isopropyl Alcohol		1 T.	15 ml
Powdered Clay or Talc		enough to make paste	

Mixing: Mix stearic acid or powdered paraffin with petroleum distillate, then add caustic soda, then denatured or isopropyl alcohol. Stir in enough powdered clay or talc to make a thick paste.

Use: With a damp cloth or sponge. Rub a small amount on the brass surface you wish to polish. Rinse off with clean, warm water.

CAUTIONS: Caustic soda heats on contact with water and can cause severe burns to skin. Handle with care. Store in an airtight container. Both denatured and isopropyl alcohol may be toxic if taken internally and both are flammable.

Notes: For proper grade of alcohol, see Appendix 4 *Denatured Alcohol.*

CHROMIUM CLEANER AND POLISH

Ingredients: Isopropyl Alcohol   to suit
           Lampblack   6 T.   84 g.

Mixing: Mix isopropyl alcohol into lampblack to form a heavy paste.

Use: Apply thin layer to chrome surface with cloth. When film forms, wipe to polish surface.

CAUTIONS: Isopropyl alcohol may be mildly toxic by ingestion and is flammable.

Notes:

COPPER CLEANER

Ingredients: Sodium Bicarbonate   1/4 C.   57 g.
           Sodium Metasilicate   1/2 C.   113 g.
           Trisodium Phosphate   1/4 C.   57 g.

Mixing: Mix all three together well.

Use: Use a small amount on a damp, soft cloth, rub well, and rinse off with warm clean water.

Notes:

COPPER-POLISHING STONE

Ingredients: Sodium Bisulfate Powder   5 oz.   140 g.
           Calcium Sulfate   8 oz.   224 g.
           Clay   16 oz.   448 g.
           Powdered Quartz   6 oz.   168 g.
           Water   to suit

Mixing: Mix sodium bisulfate powder, calcium sulfate, clay, and powdered quartz together. Add a sufficient amount of water to form a doughlike consistency. When dough is formed, press into molds, and allow to dry.

Use: This stone may be rubbed directly on copper. Or a damp cloth or sponge may be rubbed over the stone and then applied to the copper.

CAUTIONS: Sodium bisulfate is toxic in solution and an irritant to eyes and skin.

Notes:

GENERAL PURPOSE DEODORANT AND DISINFECTANT

Ingredients:			
	Soap Chips	1 lb.	454 g.
	Water	1 qt.	1 L.
	Pine Oil	1 qt.	1 L.

Mixing: Gently stir soap chips into water. If a thick layer of suds forms, stop and wait until they subside. When soap chips are thoroughly dissolved, add pine oil and mix well.

Use: Use full strength or dilute with water, depending on the job to be done.

Notes:

DISHWASHING COMPOUND

Ingredients:			
	Trisodium Phosphate	3 C.	681 g.
	Tetrasodium Pyrophosphate	1 C.	227 g.
	Borax	1/2 C.	113 g.

Mixing: Mix all three ingredients together well. Store in an airtight container.

Use: Use amount, depending on the hardness of the water. Normally, 2 tablespoons (28 g.) per sink or dishpan of water is sufficient.

CAUTIONS: Trisodium phosphate is a skin irritant and moderately toxic by ingestion. Use rubber gloves.

Notes:

ENAMEL CLEANER

Ingredients:			
	Sodium Carbonate (Soda Ash)	3 T.	42 g.
	Sodium Metaphosphate	1 T.	14 g.
	Soap Powder	2 T.	28 g.
	Fine Pumice Powder	1 C.	227 g.

Mixing: Mix all four ingredients together well.

Use: Apply with a damp rag or sponge to enamel surfaces.

Notes:

FURNITURE POLISH, LEMON OIL

Ingredients: Lemon Extract   1 T.   15 ml.
            Mineral Oil   1 qt.   1 L.

Mixing: Stir lemon extract into mineral oil. Store in glass or plastic bottle.

Use: Pour or spray a small amount directly on furniture and polish with a soft cloth. Do not get on upholstery.

Notes:

OAK FURNITURE CLEANER

Ingredients: Soda Ash   1/2 C.   113 g.
            Warm Water   1 qt.   1 L.

Mixing: Dissolve soda ash into warm water.

Use: Apply to furniture with a soft scrub brush and rinse with clean water.

Notes:

FURNITURE POLISH, OIL AND WAX

Ingredients: Mineral Oil   1 qt.   1 L.
            Carnauba Wax   2 T.   28 g.

Mixing: Put ingredients in the top of a double boiler and heat until the carnauba wax melts and can be mixed with the mineral oil. Then cool and store in glass or plastic bottles.

Use: Apply to furniture with a soft scrub brush and rinse with clean water.

Notes:

FURNITURE POLISH, SILICONE

Ingredients: Silicone Oil   2 T.   30 ml.

             Mineral Oil   1 qt.   1 L.

Mixing: Stir ingredients together. This formula does not require heating.

Use: Apply to furniture with a soft scrub brush and rinse well with clean water.

Notes:

FURNITURE POLISH, THIN FILM

Ingredients: Mineral Oil   3 C.   711 ml.

             Benzene   2 C.   474 ml.

Mixing: Stir both ingredients together.

Use: Apply to furniture with a soft scrub brush and rinse with clean water.

CAUTIONS: This is a flammable mixture (benzene is flammable). Store in closed containers and exercise due caution when using.

Notes:

FURNITURE POLISH, WATER EMULSION

Ingredients: Mineral Oil   1-1/4 C.   296 ml.

             Pine Oil (Steam Distilled)   1 T.   15 ml.

             Liquid Detergent   4 T.   60 ml.

             Water   1-1/2 C.   355 ml.

Mixing: Stir together mineral oil, pine oil, and liquid detergent. Keep stirring until the mixture is clear. Then add water very slowly in a thin stream, stirring constantly.

Use: Apply to furniture with a soft scrub brush and rinse with clean water.

Notes:

FURNITURE SCRATCH REMOVER

Ingredients:			
Ground Pecan Nut Meal	1 T	14 g.
Mineral Oil		enough to make thick paste

Mixing: Make sure the pecan nut meal is very finely ground, then add a few drops of mineral oil and mix thoroughly. Once a thick paste has formed, store in small tins.

Use: Rub the corner of a cloth in this paste and rub the cloth over any scratches you wish to cover.

Notes:

GLASS CLEANER CAKE

Ingredients:		
Granulated Soap	1/4 C.	57 g.
Hot Water	1/4 C.	59 ml.
Fuller's Earth	1/2 C.	113 g.
Calcium Carbonate (Chalk)	1/4 C.	57 g.

Mixing: Dissolve soap in hot water. Separately mix fuller's earth and chalk. Add this to the soap solution. Mix well and press into a paper-box mold, such as a large matchbox.

Use: Rub a damp cloth or sponge over the surface of the block and rub the window panes. When dry, wipe off with a dry cloth.

Notes:

GLASS CLEANER FOR GENERAL USE

Ingredients:		
Sodium Carbonate (Soda Ash)	6 oz.	168 g.
Sodium Bicarbonate (Baking Soda)	4 oz.	112 g.

Mixing: Dry-mix ingredients thoroughly and transfer to large, open-mouth jar.

Use: Dampen a cloth, dip into the mixture, and apply to glass. When it is dried to a white powder, remove with a soft, lintfree cloth.

Notes:

GLASS CLEANER, POWDER

Ingredients: Trisodium Phosphate   8 oz.   224 g.
Tetrasodium Pyrophosphate   2 oz.   56 g.

Mixing: Dry-mix ingredients and transfer to open-mouth jar or can.

Use: Dampen a cloth, dip into powder, and apply to the glass to be cleaned. Wait until a dry, white film appears on the glass and polish with a soft, lintfree cloth.

CAUTIONS: Trisodium phosphate is a skin irritant and moderately toxic by ingestion. Use rubber gloves.

Notes:

GLASS POLISH

Ingredients: Calcium Carbonate (Chalk)   1-1/4 C.   284 g.
Ground Quassia   2 T.   28 g.
Ammonium Carbonate   2 T.   28 g.

Mixing: Mix all three ingredients together well.

Use: Apply to glass with a damp cloth or sponge and rinse with water.

CAUTIONS: When ammonium carbonate is heated, irritating fumes may result.

Notes·

GLASS SCRATCH REMOVER

Ingredients: Iron Oxide (Jeweler's Rouge)   2 T.   28 g.
Glycerin   2 T.   30 ml.
Water   2 T.   30 ml.

Mixing: Mix iron oxide and glycerin, then add water slowly until a thick paste is formed.

Use: Apply with a damp cloth, which has been folded in half several times to form a pad. Rub quite hard for several minutes, or until the scratches begin to disappear. Then rinse with clean water and inspect. Rub more if needed.

Notes:

GOLD POLISH

Ingredients:			
Fuller's Earth	1 C.	227 g.
Calcium Carbonate (Chalk)	1 C.	227 g.
Ammonium Sulfate	2 T.	28 g.
Aluminum Powder	1 T.	14 g.

Mixing: Mix all ingredients together with a fork.

Use: With a damp cloth, pick up a small amount and rub on gold to clean and polish. Rinse with clean water.

Notes:

GREASE SPOT REMOVER

Ingredients:			
Sodium Aluminate	1 C.	227 g.
Water	2 qts.	2 L.

Mixing: Dissolve sodium aluminate in water by mixing with a spoon.

Use: Warm the solution and immerse the soiled area. For smaller areas, place a blotter under the spot and apply the warm solution with a sponge. Let dry and wash normally.

Notes:

INK ERADICATOR

Ingredients:			
Aluminum Potassium Sulfate (Alum)	3 T.	42 g.
Citric Acid	3 T.	42 g.
Water	5 T.	75 ml.

Mixing: Mix alum and citric acid together, then dissolve in water.

Use: Apply a few drops to ink spot and allow to stand for about 5 minutes. Use a little more if necessary.

Notes:

INK REMOVER FOR HANDS

Ingredients: Denatured Alcohol or		
Isopropyl Alcohol	3 T.	45 ml.
Glycerin	1 T.	15 ml.
Titanium Trichloride	1 T.	14 g.

Mixing: Mix the three ingredients together.

Use: Rub a small amount into stained area; wash with clear water.

CAUTIONS: Denatured or isopropyl alcohol may be toxic if taken internally and are flammable. Titanium trichloride is toxic and can be flammable. Care should be taken when opening the container; the dry powder may burn when exposed to air.

Notes: For proper grade of alcohol, see Appendix 4, *Denatured Alcohol.*

INK SPOT REMOVER

Ingredients: Sodium Perborate	1/2 t.	2 g.
Water	1/2 C.	118 ml.

Mixing: Mix sodium perborate into water.

Use: Soak spot to be cleaned, wipe with a cloth or sponge dipped in clean water, then blot-dry with paper towels.

Notes:

MAGNESIUM METAL CLEANER

Ingredients: Caustic Soda	1 T.	14 g.
Soda Ash	2 T.	28 g.
Diglycol Stearate	1/2 T.	7 g.
Water	1 qt.	1 L.

Mixing: Stir dry ingredients into water until they are all dissolved.

Use: Dip a soft cloth in this solution and rub over the metal to be cleaned. Rinse off with clean water.

CAUTIONS: Caustic soda heats on contact with water and can cause severe burns to skin. Handle with care. Store in an airtight container.

Notes:

MARBLE CLEANING POWDER

Ingredients:			
Sodium Sulfate	3/4 C.	170 g.
Sodium Sulfite	1/4 C.	57 g.

Mixing: Mix ingredients together.

Use: Apply with a damp cloth to marble surfaces.

Notes:

ALL-PURPOSE METAL CLEANER

Ingredients:			
Trisodium Phosphate	1-1/2 C.	341.0 g.
Soda Ash	6 C.	1.4 kg.
Bicarbonate of Soda (Baking Soda)	2-1/2 C.	567.0 g.

Mixing: Mix all ingredients together with fork.

Use: Apply with a damp cloth or sponge to metal surfaces and rinse with clean water.

CAUTIONS; Trisodium is a skin irritant and moderately toxic by ingestion. Use rubber gloves.

Notes:

ALL-PURPOSE METAL POLISH

Ingredients:			
Household Ammonia	1/2 C.	118 ml.
Denatured Alcohol or		
Isopropyl Alcohol	1/2 C.	118 ml.
Diatomaceous Earth	1 C.	227 g.
Water		

Mixing: Mix household ammonia with alcohol, then stir in diatomaceous earth in small amounts at a time. Add just enough water to reach a thick, creamy consistency.

Use: Shake well before each use, apply with a cloth, and rinse with clean water.

CAUTIONS: Denatured and isopropyl alcohols may be toxic by ingestion and are flammable. Avoid vapors of household ammonia.

Notes: For proper grade of alcohol, see Appendix 4, *Denatured Alcohol.*

METAL POLISHING CLOTH

Ingredients: Paraffin Oil   1 qt.   1 L.
            Petrolatum   1/2 C.   118 ml.

Mixing: Warm paraffin oil and petrolatum and stir together.

Use: Soak a square of flannel cloth in this solution and wring out the excess so that the cloth is saturated with the oil, but not dripping. Rub lightly over metal surface to be polished.

Notes:

OIL AND GREASE SPOT REMOVER

Ingredient: Fuller's Earth (or Diatomaceous Earth, from swimming-pool supplier)

Use: Sprinkle on spot, brush in well, allow to remain for a few minutes, and remove with a damp cloth.

Notes:

EXTRA-STRENGTH CLEANER FOR PAINT AND WOODWORK

Ingredients: Water   1-1/2 C.   355 ml.
            Diglycol Stearate   4 T.   56 g.
            Kerosene   1 C.   237 ml.
            Trisodium Phosphate   1 T.   14 g.

Mixing: Heat water to 150° F. Add diglycol stearate and stir until dissolved. Add kerosene and trisodium phosphate. Maintain temperature and mix vigorously with an eggbeater or electric mixer until a milky emulsion forms. Turn off heat and continue stirring until the temperature drops to 95° F. The emulsion will then be stable.

Use: Apply to painted surface with brush; remove with clean water.

CAUTIONS: Kerosene is toxic if taken internally and flammable. Trisodium phosphate is a skin irritant and moderately toxic by ingestion. Use rubber gloves.

Notes:

PAINT SPOT REMOVER

Ingredients:	Turpentine	1 C.	237 ml.
	Household Ammonia	1 C.	237 ml.

Mixing: Stir ingredients together.

Use: Rub into the spot with a cloth or sponge and wipe dry with a clean cloth.

CAUTIONS: Turpentine is toxic if taken internally and flammable. Use rubber gloves. Avoid vapors of household ammonia.

Notes:

PAINT AND WALL CLEANER I

Ingredients:	Corn Flour	1/2 C.	113.0 g.
	Copper Sulfate	1 T.	14.0 g.
	Aluminum Potassium Sulfate (Alum)	1/8 t.	0.5 g.
	Hot Water	1 qt.	1.0 L.

Mixing: Mix all ingredients together thoroughly.

Use: Apply to painted woodwork and walls; wash off with clear water.

CAUTIONS: Copper sulfate is highly toxic.

Notes:

PAINT AND WALL CLEANER II

Ingredients:	Trisodium Phosphate	32 oz.	907.0 g.
	Powdered Soap	4 oz.	112.0 g.
	Caustic Soda	2 oz.	56.0 g.
	Water, Hot	1 gal.	3.8 L.

Mixing: Dry-mix trisodium phosphate, powdered soap, and caustic soda with a wooden fork in a wooden, ceramic, or glass bowl. Store in an airtight container. Before using, add 2 ounces (56 g.) of this dry mixture to hot water and stir thoroughly.

Use: Apply to painted woodwork; wash off with clear water.

CAUTIONS: Caustic soda heats on contact with water and can cause severe burns to skin. In the event contact is made with skin, flush immediately with clear water for prolonged periods of time. Handle with care. Trisodium phosphate is an irritant to skin and moderately toxic by ingestion. Use rubber gloves.

Notes.

PINE OIL DISINFECTANT

Ingredients: Pine Oil   1 C.   237 ml.
             Liquid Soap   2 C.   474 ml.

Mixing: Mix the pine oil and soap together by stirring.

Use: Add one part of this mixture to fifty parts of water and apply wherever a disinfectant is needed.

Notes:

PLASTIC POLISH

Ingredients: Sodium Hypochlorite   3 T.   42 g.
             Caustic Soda   1/2 t.   2 g.
             Water   2 C.   474 ml.

Mixing: Carefully stir sodium hypochlorite and caustic soda into water.

Use: Apply to plastic furniture and other plastic materials with a soft cloth or a sponge. Rinse with clean water.

CAUTIONS: Sodium hypochlorite and caustic soda are toxic by ingestion or inhalation, and both are highly irritating on the skin and eyes. Use caution in handling. In event of skin contact, flush areas liberally with water. Caustic soda heats on contact with water and causes severe burns. Store in an airtight container.

Notes:

PORCELAIN CLEANER, LIQUID

Ingredients: Mineral Oil   1 C.   237 ml.
             Deodorized Kerosene   1 C.   237 ml.
             Oil-base Perfume   as desired

Mixing: Mix mineral oil and deodorized kerosene thoroughly, then add perfume while stirring.

Use: Dampen a cloth with this mixture—do not saturate it—and rub over area to be cleaned. Wipe dry and clean with either a dry corner of the original cloth, or a fresh, dry cloth.

CAUTIONS: Kerosene is toxic if taken internally and flammable.

Notes:

PORCELAIN CLEANER, POWDERED

Ingredients:			
Trisodium Phosphate	1/2 C.	113 g.
Soap Powder	1/2 C.	113 g.
Fine Chalk or Talc	1 C.	226 g.

Mixing: Dry-mix the three ingredients.

Use: Apply to porcelain with damp cloth or sponge and rinse with clear water.

CAUTIONS: Trisodium phosphate is an irritant to skin and moderately toxic by ingestion. Use rubber gloves.

Notes:

REMOVING WATER SPOTS FROM FURNITURE

Ingredients:			
Lemon Extract	10 Drops	1 ml.
Denatured Alcohol or		
Isopropyl Alcohol	2 C.	474 ml.

Mixing: Mix lemon extract into denatured or isopropyl alcohol.

Use: Dampen one corner of a soft cloth and rub the spot gently until it disappears. Then rub dry with a dry end of the cloth.

CAUTIONS: Isopropyl or denatured alcohol may be toxic by ingestion and are flammable.

Notes: For proper grade of alcohol, see Appendix 4, *Denatured Alcohol.*

RUG CLEANING LIQUID

Ingredients:			
Denatured Alcohol or		
Isopropyl Alcohol	1 C.	237.0 ml.
Vinegar	5 C.	1.2 L.
Lauryl Pyridinium Chloride	1/4 t.	1.0 g.

Mixing: Mix alcohol and vinegar, then stir in lauryl pyridinium chloride. Store in bottles.

Use: Brush mixture into the rug, allow to dry, and vacuum up.

CAUTIONS: Isopropyl and denatured alcohol may be toxic by ingestion and are flammable. Lauryl pyridinium chloride may be mildly irritating to skin.

Notes: For proper grade of alcohol, see Appendix 4, *Denatured Alcohol.*

RUG DRY-CLEANING FLUID

Ingredients:			
Stoddard Solvent	1-1/2 C.	355.0 ml.
Diglycol Laurate	1/8 t.	0.6 ml.
Water	1/2 C.	118.0 ml.
Diatomaceous Earth	3 C.	681 g.

Mixing: Mix stoddard solvent, diglycol laurate, and water into diatomaceous earth.

Use: Apply a generous amount to the area to be cleaned, brush in, let dry, and vacuum.

CAUTIONS: Stoddard solvent is slightly flammable.

Notes:

SOAPLESS RUG CLEANER

Ingredients:			
White Vinegar	1 C.	237 ml.
Denatured Alcohol or		
Isopropyl Alcohol	1 qt.	1 L.

Mixing: Stir both ingredients together.

Use: Apply to rug with a sponge-head mop. Rinse with clean water.

CAUTIONS: Isopropyl and denatured alcohol may be toxic by ingestion and are flammable.

Notes: For proper grade of alcohol, see Appendix 4, *Denatured Alcohol.*

SILVER ANTITARNISH BAGS

Ingredients:			
Cadmium Acetate	1/4 C.	57 g.
Water	1 qt.	1 L.
Flannel Cloth		

Mixing: Stir cadmium acetate into water until dissolved. Soak flannel cloth in this solution and hang up to dry.

Use: When cloth is dry, either sew into a bag or else wrap silver in this treated cloth to prevent tarnishing.

CAUTIONS: Cadmium acetate is highly toxic in concentrated form.

Notes:

SILVER CLEANER AND POLISH

Ingredients:			
Water		1-1/2 C.	355 ml.
Stearic Acid		2 T.	28 g.
Soda Ash		1/2 t.	2 g.
Trisodium Phosphate		1/2 t.	2 g.
Diatomaceous Earth		1 C.	227 g.

Mixing: Heat water and stearic acid in the top of a double boiler until the acid melts. Turn off the heat and add soda ash, trisodium phosphate, and diatomaceous earth. Stir into a creamy paste. Cool and store in glass or plastic jars.

Use: Rub silver with this paste using a soft cloth. Rinse silver off with clean, warm water.

CAUTIONS: Trisodium phosphate is an irritant to skin and moderately toxic by ingestion. Use rubber gloves.

Notes:

STOVE CLEANER AND POLISH

Ingredients:			
Trisodium Phosphate		1/4 C.	57 g.
Soda Ash		1/4 C.	57 g.
Sodium Perborate		1/2 C.	113 g.

Mixing: Mix three ingredients together.

Use: Dip damp cloth or sponge into mixture, rub over area to be cleaned, and rinse with cloth rung-out in clear water.

CAUTIONS: Trisodium phosphate is an irritant to skin and moderately toxic by ingestion. Use rubber gloves.

Notes:

TOILET BOWL CLEANER I

Ingredients:	Caustic Soda	1-1/2 C.	340 g.
	Powdered Alum	3/4 C.	170 g.
	Sodium Chloride	1/4 C.	57 g.

Mixing: Mix all together and store in an airtight container.

Use: Flush toilet to get sides wet, then sprinkle mixture in bowl. Allow to stand for at least 10 minutes and scrub out with a brush and flush once more.

CAUTIONS: Caustic soda heats on contact with water and can cause severe skin burns. Handle carefully. If a burn or stinging sensation does occur, flush for several minutes with clear water. Store in an airtight container.

Notes:

TOILET BOWL CLEANER II

Ingredients:	Caustic Soda	3/4 C.	170 g.
	Sodium Bicarbonate (Baking Soda)	4 C.	908 g.

Mixing: Mix ingredients together and store in an airtight container.

Use: Flush toilet to get sides wet, then sprinkle mixture in bowl. Allow to stand for at least 10 minutes and scrub out with a brush and flush once more.

CAUTIONS: Caustic soda heats on contact with water and can cause severe burns to skin. Handle with care. If burn or stinging sensation does occur, flush for several minutes with clear water. Store in an airtight container.

Notes:

TYPE CLEANER

Ingredients:	Denatured Alcohol, or		
	Isopropyl Alcohol	1-1/2 C.	355 ml.
	Water	1 C.	237 ml.

Mixing: Stir together. Store in airtight containers.

Use: Saturate a small portion of a clean cloth with this mixture and rub over the type. Clean and dry with a dry end of the same cloth.

CAUTIONS: Isopropyl and denatured alcohol may be toxic by ingestion and are flammable.

Notes: For proper grade of alcohol, see Appendix 4, *Denatured Alcohol.*

UPHOLSTERY CLEANER

Ingredients:		
Oil Soap	1/4 C	59.0 ml.
Borax	1 T.	14.0 g.
Glycerin	1/4 C.	59.0 ml.
Ethylene Chloride	2 T.	30.0 ml.
Hot Water	3 qt.	2.8 L.

Mixing: Add oil soap to hot water and stir gently. Then add borax, glycerin, and ethylene chloride, mixing well. Cool and store in bottles.

Use: Apply a small amount with a sponge and rinse with clean water.

CAUTIONS: Ethylene chloride is toxic by ingestion, inhalation, and may be absorbed through the skin. It is also an irritant to skin or eyes. Use due caution when handling.

Notes:

VINYL CLEANER

Ingredients:		
Calcium Carbonate (Chalk)	1 C.	227 g.
Bicarbonate of Soda (Baking Soda)	3 C.	681 g.

Mixing: Stir together.

Use: Sprinkle on a damp cloth or sponge and rub on the area to be cleaned. Rinse with clean water.

Notes:

WALL CLEANER

Ingredients:		
Soda Ash	1 C.	227 g.
Ammonium Sulfate	1 T.	15 g.

Mixing: Stir ingredients together and store in container.

Use: Use 2 tablespoons (28 g.) of mixture per 1 quart (1 L.) water.

Notes:

WALLPAPER CLEANER

Ingredients:	Wheat or Corn Flour	14 oz.	392 g.
	Copper Sulfate	1-1/2 oz.	42 g.
	Powdered Alum		
	(Aluminum Potassium Sulfate)	1/4 oz.	7 g.
	Boiling Water	1 gal.	4 L.

Mixing: Mix chemicals into boiling water. Allow to cool.

Use: Apply with brush or cloth to surface and remove with a damp sponge or cloth.

CAUTIONS: Copper sulfate is highly toxic by inhalation and ingestion. Use with care.

Notes:

WINDOW CLEANING POWDER

Ingredients:	Borax	1-3/4 C.	397 g.
	Sodium Metaphosphate	3 T.	42 g.

Mixing: Mix borax and sodium metaphosphate.

Use: Pick up a small amount with a damp cloth or sponge and wipe windows. Rinse off with clean water.

Notes:

WINDOW CLEANING SPRAY I

Ingredients:	Ethylene Glycol	2 T.	30 ml.
	Water	3 C.	711 ml.

Mixing: Stir ethylene glycol into water.

Use: Spray on windows and wipe with squeegee or lintfree cloth.

CAUTIONS: Family pets love this because it is sweet. It turns to oxalic acid in their bodies and can kill them. Keep pets away when using mixture, and dispose of excess carefully.

Notes:

WINDOW CLEANING SPRAY II

Ingredients: Liquid Detergent	1-1/2 t.	7.5 ml.
Water	1 gal.	3.8 L.
Denatured Alcohol or		
Isopropyl Alcohol	5 T.	75.0 ml.

Mixing: Gently stir the liquid detergent into water; then add denatured or isopropyl alcohol. Store in bottles.

Use: Spray or sponge liberally on windows and wipe off with a squeegee or cloth.

CAUTIONS: Isopropyl and denatured alcohol may be toxic by ingestion and are flammable.

Notes: For proper grade of alcohol, see Appendix 4, *Denatured Alcohol.*

WINDOW CLEANING SPRAY III

Ingredients: Denatured Alcohol or		
Isopropyl Alcohol	1 C.	237.0 ml.
Water	2 C.	474.0 ml.
Lactic Acid	5 Drops	0.5 ml.

Mixing: Mix denatured or isopropyl alcohol with water; then stir in lactic acid. A drop or two of regular laundry bluing may be added for blue color if desired.

Use: Spray on windows, mirrors, and so on.

CAUTIONS: Denatured and isopropyl alcohol may be toxic by ingestion and are flammable.

Notes: For proper grade of alcohol, see Appendix 4, *Denatured Alcohol.*

WOODWORK CLEANER

Ingredients: Oleic Acid	1-1/2 C.	355 ml.
Triethanolamine	1/2 C.	118 ml.

Mixing: Mix chemicals together and store in an airtight container.

Use: Use 2 tablespoons (30 ml.) to 1 gallon (3.8 L.) of warm water, and apply with a cloth or sponge.

CAUTIONS: Triethanolamine may be somewhat irritating to skin and mucous membranes.

Notes:

BLEACH

Ingredients:	Chlorinated Lime	2 C.	454.0 g.
	Soda Ash (Sodium Carbonate)	3 C.	681.0 g.
	Water	1 gal.	3.8 L.

Mixing: Mix chlorinated lime and sodium ash into water, stirring well. Allow to stand for 24 hours and strain into bottles.

Use: Use 1/4 cup (57 g.) per average load of white laundry.

CAUTIONS: Chlorinated lime forms chlorine when mixed with water and is a skin irritant.

Notes:

BLUING

Ingredients:	Ultramarine Blue	3-3/4 C.	851 g.
	Sodium Bicarbonate (Baking Soda)	2-1/2 C.	567 g.
	Corn Syrup	3/4 C.	177 ml.

Mixing: Stir ultramarine blue and sodium bicarbonate into corn syrup until dissolved.

Use: Mix small amount, about 1/2 teaspoon (2 g.), in the wash load to help brighten clothes.

Notes:

ENZYME-ACTIVE DETERGENT

Ingredients:			
Pancreatin	1/4 C.	57 g.
Sodium Chloride (Salt)	1/4 C.	57 g.
Soda Ash (Sodium Carbonate)	3-1/2 C.	794 g.
Sodium Alginate	1 C.	227 g.

Mixing: Mix ingredients thoroughly and store in any large container with a lid.

Use: Use 1 cup (227 g.) in average top-loading machine.

Notes:

LOW SUDSING DETERGENT

Ingredients:			
Lauryl Pyridinium Chloride	3-3/4 C.	851.0 g.
Sodium Dodecylbenzene Sulfate	2-1/2 C.	567.0 g.
Sodium Tripolyphosphate	10-1/2 C.	2.4 kg.
Sodium Bicarbonate (Baking Soda)	8 C.	1.8 kg.

Mixing: Mix ingredients thoroughly and store in any large container with a lid.

Use: Use 1 cup (227 g.) in average top-loading machine.

CAUTIONS: Lauryl pyridinium chloride may be mildly irritating to skin.

Notes:

DRY-CLEANING FLUID

Ingredient:			
Stoddard Solvent	1 gal.	3.8 L.

Use: Place garments to be cleaned in a shallow pan, cover with fluid, and agitate gently, for about 5 minutes. Wring out and air-dry.

CAUTIONS: Stoddard solvent is moderately flammable; exercise caution.

Notes:

FABRIC SOFTENER

Ingredients:			
Denatured Alcohol	1 C.	237 ml.
Water	1 C.	237 ml.
Lauryl Pyridinium Chloride	4 C.	908 g.

Mixing: Mix denatured alcohol and water, and dissolve lauryl pyridinium chloride in this mixture.

Use: Add 2 tablespoons (28 g.) to rinse water.

CAUTIONS: Lauryl pyridinium chloride may be mildly irritating to skin. Denatured alcohol may be toxic by ingestion and is flammable.

Notes: For proper grade of alcohol, see Appendix 4, *Denatured Alcohol.*

LAUNDRY GLOSS

Ingredients:			
Acacia (Gum Arabic)	1 T.	14 g.
Water	1 pt.	474 ml.
Borax	1 oz.	28 g.
Glycerin	1 T.	15 ml.

Mixing: Soak acacia in water overnight. Add borax, stir and bring to boil. Remove from heat and add glycerin; cool and strain.

Use: Add 1 tablespoon (15 ml.) to 1 pint (473 ml.) of starch. This improves the starch and provides a smoother finish when ironing.

Notes:

LAUNDRY HELPER

Ingredients:			
Mason Sand	5 C.	1.1 kg.
Soda Ash	3 C.	681.0 g.

Mixing: Stir ingredients together.

Use: Use 1 or 2 tablespoons (14 or 28 g.) per wash load. The gentle scrubbing action of these solid ingredients greatly increases the cleaning action of soap or detergent.

Notes:

IRONING AID

Ingredients: Silicone Oil Emulsion   1 t.   5 ml.
Distilled Water   1 qt.   1 L.

Mixing: Stir silicone oil emulsion into distilled water.

Use: Spray a fine mist onto a cold iron. Then heat the iron until the water steams off the bottom. The iron is then ready for use.

Notes:

MOLD STAIN REMOVER

Ingredients: Household Ammonia   1 t.   5 ml.
Hydrogen Peroxide   4 T.   60 ml.
Water   3/4 C.   177 ml.

Mixing: Mix household ammonia and hydrogen peroxide into water.

Use: Thoroughly wet the mold-stained spot with this mixture and allow to soak for 5 minutes. Rinse well with clean water and repeat if necessary.

CAUTIONS: Avoid vapors of household ammonia. Hydrogen peroxide is highly toxic in concentrated form; relatively low toxicity in dilute aqua solution.

Notes:

STARCH

Ingredients: Corn Starch   1 C.   227 g.
Wheat Starch   1/2 C.   113 g.

Mixing: Stir ingredients together.

Use: Dissolve 2 teaspoons (8 g.) in 1 cup (237 ml.) of water. Sprinkle or spray on clothing before ironing. Proportions can be adjusted to suit different requirements or tastes.

Notes:

WATER SOFTENER

Ingredients: Soda Ash	3 C.	681.0 g.
Sodium Silicate (Waterglass)	6 C.	1.4 L.

Mixing: Stir soda ash into sodium silicate.

Use: This is a concentrated mixture. Use about 1/2 teaspoon (2 g.) for 5 gallons (19 L.) of water, depending on the hardness. You will have to experiment a little to find the right proportion for your particular water.

CAUTIONS: Sodium silicate may be irritating and caustic to skin and mucous membranes.

Notes:

ANIMAL SOIL STAIN REMOVER

Ingredients: Sodium Perborate	1 T.	14 g.
Soap Powder	1 T.	14 g.
Water	1 C.	237 ml.

Mixing: Dissolve sodium perborate and soap powder into water.

Use: Soak stained area for 5 minutes, then wipe off with damp sponge and wash normally.

Notes:

COFFEE AND ALCOHOL STAIN REMOVER

Ingredients: Glycerin	1 T.	15 ml.
Denatured Alcohol or		
Isopropyl Alcohol	1 T.	15 ml.
Water	3 T.	45 ml.
Ammonium Chloride	1 T.	14 g.

Mixing: Mix glycerin, denatured or isopropyl alcohol, and water, then add ammonium chloride and stir until dissolved.

Use: Saturate spot with mixture and let stand for 5 minutes, then rinse well with clean water.

CAUTIONS: Denatured and isopropyl alcohol may be toxic by ingestion and are flammable.

Notes: For proper grade of alcohol, see Appendix 4, *Denatured Alcohol.*

EGG-YOLK STAIN REMOVER

Ingredients:			
Glycerin		1 T.	15 ml.
Denatured Alcohol or			
Isopropyl Alcohol		1/4 C.	59 ml.
Powdered Soap		1/4 C.	57 g.

Mixing: Combine the three ingredients.

Use: Rub the mixture into stained area and allow to stand for 5 minutes, then rinse with clean water.

CAUTIONS: Denatured and isopropyl alcohol may be toxic by ingestion and are flammable.

Notes: For proper grade of alcohol, see Appendix 4, *Denatured Alcohol.*

GRASS STAIN REMOVER

Ingredient: Denatured Alcohol or Isopropyl Alcohol

Use: Wash material with grass stain, using denatured alcohol, then rinse well with clean water.

CAUTIONS: Denatured and isopropyl alcohol may be toxic by ingestion and are flammable.

Notes: For proper grade of alcohol, see Appendix 4, *Denatured Alcohol.*

INK STAIN REMOVER

Ingredients:		
Ammonium Hydroxide	1 T.	15 ml.
Hydrogen Peroxide	1 T.	15 ml.

Mixing: Mix ingredients together.

Use: Soak stain with mixture, allow to stand for 5 minutes, and rinse well with clean water.

CAUTIONS: Hydrogen peroxide is highly toxic in concentrated form; relatively low toxicity in dilute aqua solution.

Notes:

IRONING SCORCH STAIN REMOVER

Ingredients: Hydrogen Peroxide   1/2 C.   118 ml.
Water   1/2 C.   118 ml.

Mixing: Stir ingredients together.

Use: Soak scorched spot with mixture for 10 or 15 minutes, then soak in clean water for 45 minutes.

CAUTIONS: Hydrogen peroxide is highly toxic in concentrated form; relatively low in dilute aqua solution.

Notes:

MILDEW STAIN REMOVER

Ingredients: Hydrogen Peroxide   4 T.   60 ml.
Ammonium Chloride   1 T.   15 ml.
Denatured Alcohol or
     Isopropyl Alcohol   2-1/2 T.   37 ml.
Water   1 C.   237 ml.

Mixing: Mix hydrogen peroxide, ammonium chloride, denatured or isopropyl alcohol into water.

Use: Soak stained area in mixture for 5 minutes, rinse well with clean water, and repeat if necessary.

CAUTIONS: Hydrogen peroxide is highly toxic in concentrated form; relatively low toxicity in dilute aqua solution. Isopropyl and denatured alcohol may be toxic by ingestion and are flammable.

Notes: For proper grade of alcohol, see Appendix 4, *Denatured Alcohol.*

PERSPIRATION STAIN REMOVER

Ingredients: Sodium Perborate   2 T.   30 g.
Water   2 C.   474 ml.

Mixing: Dissolve sodium perborate in water.

Use: Test on small area of fabric for color fastness before applying to stained area. After stain is removed, rinse with clean water.

Notes:

RUST STAIN REMOVER

Ingredients: Potassium Persulfate   1-1/2 T.   21 g.
Water   1 pt.   474 ml.

Mixing: Dissolve potassium persulfate in water.

Use: Saturate spot with mixture for 5 minutes, then rinse well with clean water.

CAUTIONS: Potassium persulfate is moderately toxic.

Notes:

URINE STAIN REMOVER

Ingredient: 10% Citric Acid Solution

Use: Saturate stained area with solution for 5 minutes, then rinse well with clean water.

Notes:

"I can't explain now, but don't take your jacket off at work."

In those formulas in which it is felt that there is a possibility of allergic reaction, it is suggested that the formulas be applied to a small test area to determine if an allergic reaction takes place and if it does, use should be discontinued. However, if the reader or user of the formula is prone to allergic reactions, all personal products should be tested in this way.

It is also suggested that all containers and mixing utensils for personal care formulas, especially those used near the eyes, should be sterilized or at least thoroughly washed.

ACNE LOTION

Ingredients:		
Isopropyl Alcohol	3 T.	45.0 ml.
Distilled Water	6 T.	90.0 ml.
Glycerin	1 t.	5.0 ml.
Triethanolamine	1/8 t.	0.6 ml.

Mixing: Stir all ingredients together.

Use: Apply to problem areas with a cotton swab.

CAUTIONS: Isopropyl alcohol may be toxic by ingestion and is flammable. The triethanolamine may be somewhat irritating to skin and mucous membranes.

Notes: For discussion on grades of alcohol, see Appendix 4, *Denatured Alcohol.*

ADHESIVE TAPE REMOVAL FROM SKIN

Ingredient: Methyl Salicylate (Oil of Wintergreen)

Use: Apply a small amount of oil to area where tape adheres to skin and pull gently. The tape will come off easily and painlessly.

CAUTIONS: Methyl salicylate is highly toxic by ingestion in concentrated form.

Notes:

AFTER-SHAVE LOTION

Ingredients:			
Boric Acid	1 t.	4 g.
Glycerin	2 T.	30 ml.
Water	2 C.	474 ml.
Water-base Perfume	to suit	
Isopropyl Alcohol	1 C.	237 ml.

Mixing: Stir boric acid, glycerin and perfume into water, then stir in isopropyl alcohol.

Use: Splash on face after shaving.

CAUTIONS: Isopropyl alcohol may be toxic by ingestion and is flammable.

Notes: For discussion on grades of alcohol, see Appendix 4, *Denatured Alcohol.*

"Want me to get you a shopping cart?"

AFTER-SHAVE FACE TALC

Ingredients:			
Talc		2 C.	454 g.
Boric Acid		1 T.	14 g.
Magnesium Stearate		5 t.	20 g.
Zinc Oxide		1 T.	14 g.
Oil-base Perfume		to suit	

Mixing: Mix ingredients together with a fork and sift through a flour sifter.

Use: Apply after shaving as a soothing dust-on talc.

CAUTIONS: Zinc oxide is poisonous if taken internally.

Notes:

ANALGESIC SALVE

Ingredients:			
Anhydrous Lanolin		1-3/4 C.	398 g.
Menthol		1/4 C.	57 g.
Methyl Salicylate (Wintergreen Oil)		1/4 C.	59 ml.

Mixing: Heat anhydrous lanolin in the top of a double boiler until it melts, then stir in menthol and methyl salicylate. Cool until it just begins to solidify, then put in jars. Allow to cool, room temperature, uncovered and cover when cool.

Use: Use a small amount to rub into sore muscles.

CAUTIONS: Methyl salicylate is highly toxic by ingestion in concentrated form.

Notes:

ANTACID LIQUID

Ingredients:			
Sodium Bicarbonate (Baking Soda)		1 T.	14 g.
Sugar		1 T.	14 g.
Peppermint Oil		10 Drops	1 ml.
Water		2 C.	474 ml.

Mixing: Mix sodium bicarbonate, sugar, and peppermint oil into water.

Use: Take 1 or 2 teaspoons (5 to 10 ml.) to relieve indigestion.

Notes:

ANTICHAP STICK FOR LIPS

Ingredients: Beeswax 1/4 C. 57 g.
Castor Oil 1/4 C. 59 ml.
Sesame Oil 3 T. 45 ml.
Anhydrous Lanolin 2 T. 28 g.

Mixing: Melt beeswax in the top of a double boiler. When melted, stir in remaining ingredients. Partially cool and pour into small molds.

Use: Rub on lips to protect them from harsh weather.

Notes: The molds for this and the following formula can be made from small match boxes. (Lipstick does not have to be round.)

ANTICHAP STICK FOR LIPS II

Ingredients: Beeswax 4 oz. 112 g.
Castor Oil 3 oz. 84 g.
Cocoa Butter 2 oz. 56 g.
Anhydrous Lanolin 1/2 oz. 14 g.
Mineral Oil 1/2 oz. 14 g.

Mixing: Melt all ingredients together in the top of a double boiler over low heat. When thoroughly mixed, allow the mass to cool down to just above the solidification point and pour into suitable molds.

Use: Rub on lips to protect them from harsh weather.

Notes:

ANTIPERSPIRANT LIQUID I

Ingredients: Oxyquinoline Sulfate 1 T. 14 g.
Water 1 pt. 434 ml.

Mixing: Mix the oxyquinoline sulfate into water until dissolved.

Use: This liquid may be applied with a plastic spray bottle, ball of cotton, or flannel pads.

CAUTIONS: Oxyquinoline sulfate is moderately toxic in concentrated form.

Notes:

ANTIPERSPIRANT LIQUID II

Ingredients:			
Isopropyl Alcohol		1/2 C.	118 ml.
Water		2-1/2 C.	592 ml.
Powdered Alum (Aluminum Potassium Sulfate)		1 T.	14 g.
Zinc Oxide, Powdered		1 T.	14 g.

Mixing: Mix isopropyl alcohol and water, then stir in powdered alum and zinc oxide. Store in an airtight container.

Use: Dab on with cotton swab, or put in plastic spray bottle and spray on.

CAUTIONS: Isopropyl alcohol may be toxic by ingestion and is flammable. Zinc oxide is poisonous if taken internally.

Notes: For discussion on grades of alcohol, see Appendix 4, *Denatured Alcohol.*

ANTIPERSPIRANT LIQUID III

Ingredients:		
Aluminum Potassium Sulfate (Powdered Alum)	3 T.	42 g
Water	2 C.	474 ml.

Mixing: Stir aluminum potassium sulfate into water.

Use: Apply to underarms with a plastic spray bottle, ball of cotton, or flannel pads.

Notes:

ANTIPERSPIRANT POWDER I

Ingredients:		
Oxyquinoline Sulfate	1-1/4 C.	284 g.
Talc	2 T.	28 g.

Mixing: Mix ingredients together with a fork.

Use: Dust on or the powder may be sprayed from a plastic squeeze bottle if the bottle is only filled half-full.

CAUTIONS: Oxyquinoline sulfate is moderately toxic in concentrated form.

Notes:

ANTIPERSPIRANT POWDER II

Ingredients: Aluminum Potassium Sulfate (Alum)   1 T.   14 g.
Boric Acid   1/4 C.   57 g.
Talc   1/3 C.   76 g.

Mixing: Mix the ingredients together with a fork.

Use: Dust on or the powder may be sprayed from a plastic squeeze bottle if the bottle is only filled half-full.

Notes:

ANTISEPTIC FOR POISON IVY AND OAK RASH

Ingredients: Water   4 oz.   120 ml.
Isopropyl Alcohol   6 oz.   180 ml.
Ferric Chloride   1/2 oz.   14 g.

Mixing: Mix water and isopropyl alcohol and then add ferric chloride while stirring.

Use: Wash affected areas with this solution.

CAUTIONS: Isopropyl alcohol may be moderately toxic by ingestion and is flammable. Ferric chloride is mildly toxic.

Notes: For discussion on grades of alcohol, see Appendix 4, *Denatured Alcohol.*

ASTRINGENT SKIN CREAM

Ingredients: Mineral Oil   2 C.   474 ml.
White Beeswax   1/2 C.   113 g.
Water   1 C.   237 ml.
Borax   1-1/2 T.   21 g.
Aluminum Potassium Sulfate
(Powdered Alum)   2 T.   28 g.

Mixing: Heat mineral oil and white beeswax in the top of a double boiler until the beeswax melts. Stir to mix the two and cool down to 120° F. In a separate pan, heat water to 120° F. and slowly stir in borax and aluminum potassium sulfate. Then slowly pour the second mixture into the first, stirring gently all the while. Cool until it just starts to solidify, pour into jars, cool to room temperature, and cover.

Use: Use this cream to smooth and tighten skin around the neck, elbows, wrists, and ankles.

Notes:

ASTRINGENT SKIN LOTION

Ingredients:			
Water		3 C.	711 ml.
Glycerin		1 T.	15 ml.
Aluminum Potassium Sulfate (Alum)		2 T.	28 g.

Mixing: Mix water and glycerin, then add aluminum potassium sulfate, stirring until it is dissolved.

Use: A refreshing astringent.

Notes:

ASTRINGENT LOTION, GENERAL PURPOSE

Ingredients:			
Vodka		2 oz.	60 ml.
Glycerin		1/4 oz.	7 ml.
Water		12 oz.	360 ml.
Aluminum Potassium Sulfate (Alum)		1/2 oz.	14 g.
Water-soluble Perfume		to suit	

Mixing: Mix vodka and glycerin into water and stir. Add aluminum potassium sulfate to solution; then add perfume, mixing thoroughly.

Use: Apply to body with cotton balls or pads.

Notes:

ATHLETE'S FOOT OINTMENT

Ingredients:			
Anhydrous Lanolin		1 C.	227 g.
Flowers of Sulfur		1 T.	14 g.

Mixing: Heat anhydrous lanolin in the top of a double boiler until it melts, then stir in flowers of sulfur, mixing well. Allow to cool until it just starts to solidify, then pour into jars, and cool to room temperature.

Use: Rub a small amount on infected areas of feet.

Notes:

ATHLETE'S FOOT POWDER

Ingredients:			
Sodium Thiosulfate	1/3 C.	76 g.
Boric Acid	1 C.	227 g.
Talc	2/3 C.	151 g.

Mixing: Stir ingredients together with a fork.

Use: Sprinkle on feet, being sure to get between the toes.

Notes:

BABY OIL I

Ingredients:			
Mineral Oil, White U.S.P.	1 pt.	474 ml.
Oil-base Perfume	to suit	

Mixing: Mix an oil-base perfume into the white mineral oil.

Use: Apply sparingly.

Notes:

BABY OIL II

Ingredients:			
Almond Oil	2 T.	30 ml.
Olive Oil	8 T.	120 ml.
Oil-base Perfume	to suit	

Mixing: Mix almond oil and olive oil, and oil-base perfume as desired.

Use: Apply sparingly.

CAUTIONS: Almond oil vapors are toxic.

Notes:

BABY POWDER

Ingredients: Talc, U.S.P. Grade   1 lb.   454 g.
            Oil-base Perfume   to suit

Mixing: A flour sifter is an excellent way to distribute the perfume evenly into the talc.

Use: Use for dusting the baby after bathing.

Notes:

AFTER-BATH TALC

Ingredients:		
Talc	1 C.	227 g.
Zinc Stearate	2 T.	28 g.
Boric Acid	1 T.	14 g.
Magnesium Carbonate	1 T.	14 g.
Oil-base Perfume	to suit	

Mixing: Mix ingredients together with a fork and sift with a flour sifter.

Use: Use as a body talc.

Notes:

RHEUMATIC AND ARTHRITIC BATH SALTS

Ingredients:		
Sodium Sulfate	1 C.	227 g.
Sodium Chloride (Salt)	1 C.	227 g.
Sodium Bicarbonate (Baking Soda)	4 C.	908 g.

Mixing: Mix ingredients together. Store in jar or plastic container.

Use: Use about 1/2 cup (113 g.) per tub of hot water.

Notes:

FOAMING BATH SALTS

Ingredients:			
Sodium Bicarbonate	9 T.	126 g.
Tartaric Acid (Cream of Tartar)	7-1/2 T.	105 g.
Cornstarch	2 T.	28 g.

Mixing: Stir all the ingredients together with a spoon or fork and store in jar or plastic container.

Use: Use about 2 tablespoons (28 g.) in a tub of water.

Notes:

SOFTENING BATH SALTS

Ingredient: Sodium Sesquicarbonate	1 lb.	454 g.

Use: Add about 2 tablespoons (28 g.) to bath water.

Notes:

BAY RUM AFTER-SHAVE LOTION

Ingredients:		
Orange Oil	1 t.	5.0 ml.
Isopropyl Alcohol	2 C.	474.0 ml.
Citric Acid	1/8 t.	0.6 ml.
Glycerin	1 T.	15.0 ml.
Water	2 C.	474.0 ml.
Talc	1 T.	14.0 g.

Mixing: Dissolve orange oil in isopropyl alcohol, then add citric acid and glycerin, stirring well. Add water and finally stir in talc.

Use: Use as an after-shave lotion.

CAUTIONS: Isopropyl alcohol may be moderately toxic by ingestion and is flammable.

Notes: The juice of one lemon may be used in place of the citric acid. For discussion on grades of alcohol, see Appendix 4, *Denatured Alcohol.*

BEAUTY CLAY

Ingredients: | | | |
|---|---|---|
| Clay, Powdered | 5 lb. | 2.3 kg. |
| Tincture of Benzoin | 1/4 C. | 59.0 ml. |
| Water-base Perfume | to suit | |
| Water | | |

Mixing: Use enough water with the remaining ingredients to make a thick paste.

Use: Cover face and neck with clay, let dry thoroughly—about 20 to 30 minutes—then wash off with cold water.

Notes:

BEAUTY MASK

Ingredients: | | | |
|---|---|---|
| Clay | 8 C. | 2 kg. |
| Talc | 1-3/4 C. | 398 g. |
| Titanium Dioxide | 1/4 C. | 57 g. |

Mixing: Mix clay, talc, and titanium dioxide. Then add water to form a paint-like consistency. Apply with a soft brush to face area. Allow to dry. After 30 minutes, wash off with clear water.

Notes:

BEAUTY MILK

Ingredients: | | | |
|---|---|---|
| Mineral Oil | 1/2 C. | 118 ml. |
| Lemon Extract | 2 T. | 30 ml. |
| Cetyl Alcohol | 1 T. | 14 g. |
| Triethanolamine | 2 T. | 30 ml. |
| Glycerin | 4 T. | 60 ml. |
| Oil-base Perfume | to suit | |
| Water | 3 C. | 711 ml. |

Mixing: Mix mineral oil, lemon extract, cetyl alcohol, triethanolamine, and glycerin. Then add this mixture slowly to water, stirring constantly. Mix in desired amount of an oil-base perfume.

Use: Apply liberally to skin, especially dried areas.

CAUTIONS: Triethanolamine may be somewhat irritating to skin and mucous membranes.

Notes:

BODY-MASSAGE LOTION

Ingredients:			
Menthol Crystals		1/8 t.	0.5 g.
Isopropyl Alcohol		2 T.	30.0 ml.
Glycerin		2 T.	30.0 ml.
Water		1 pt.	474.0 ml.

Mixing: Dissolve menthol crystals in isopropyl alcohol and add glycerin while stirring. Stir this solution into water until it is completely homogenous.

Use: Use as any massage lotion would be used.

CAUTIONS: Isopropyl alcohol may be moderately toxic by ingestion and is flammable.

Notes: For discussion on grades of alcohol, see Appendix 4, *Denatured Alcohol.*

BLACKHEAD REMOVER

Ingredients:			
Isopropyl Alcohol		1-3/4 C.	474 ml.
Water		3 C.	711 ml.
Triethanolamine		1 t.	5 ml.
Glycerin		2 T.	30 ml.

Mixing: Mix isopropyl alcohol into water, then, stirring, add triethanolamine and glycerin.

Use: Apply to skin at night. Wipe or wash off in the morning.

CAUTIONS: Triethanolamine may be somewhat irritating to skin and mucous membranes. Isopropyl alcohol may be toxic by ingestion and is flammable.

Notes: For discussion on grades of alcohol, see Appendix 4, *Denatured Alcohol.*

BRILLIANTINE

Ingredients:			
Glycerin		1 C.	237 ml.
Denatured Alcohol		1 C.	237 ml.
Perfume (Water-soluble)		to suit	

Mixing: Mix glycerin into denatured alcohol and add perfume.

Use: Apply to hair and comb.

CAUTIONS: Denatured alcohol may be toxic by ingestion and is flammable.

Notes: For proper grade of alcohol, see Appendix 4, *Denatured Alcohol.*

BUBBLE BATH

Ingredients:			
Sodium Lauryl Sulfate	1-3/4 C.	398 g.
Sodium Sesquicarbonate	3 C.	681 g.
Sodium Alginate	1 t.	4 g.
Water-base Perfume	to suit	

Mixing: Mix ingredients, adding a few drops of perfume.

Use: Use in bath water as desired.

Notes:

BURN TREATMENT

The latest medical information on burn treatment as used in hospital burn centers indicates that the most effective method is to flush the burned area with water at tap temperature for prolonged periods of time, half an hour or more, at least. Salves and ointments are not recommended. Serious and extensive burns should receive prompt medical treatment.

Notes:

CALLOUS SOFTENER AND REMOVER

Ingredients:			
Castor Oil	1/2 C.	118 ml.
Beeswax	1/2 C.	113 g.
White Soap	1 T.	14 g.
Sodium Thiosulfate	1 t.	4 g.

Mixing: Melt castor oil, beeswax, and white soap together in the top of a double boiler. Add sodium thiosulfate and cool.

Use: Apply a small amount to callous and cover with a light bandage overnight. Wash with hot water in the morning.

Notes:

COMB AND HAIRBRUSH CLEANER

Ingredients: Household Ammonia   1 T.   15 ml.
            Water   2 C.   474 ml.

Mixing: Mix ingredients together.

Use: Soak combs and brushes for 10 minutes or so and then rinse with clean water.

CAUTIONS: Avoid the vapors of household ammonia.

Notes:

COMPLEXION MOISTURIZER

Ingredients: Light Mineral Oil   1 C.   237 ml.
            Oil-base Perfume   to suit

Mixing: Stir a few drops of perfume into mineral oil.

Use: Use sparingly; wipe off excess with a tissue.

Notes:

CONTACT LENS CLEANING FLUID

Ingredients: Sodium Bicarbonate (Baking Soda)   1/8 t.   0.5 g.
            Sodium Chloride (Salt)   1/8 t.   0.5 g.
            Distilled Water   1/2 C.   118.0 ml.

Mixing: Dissolve sodium bicarbonate and sodium chloride in distilled water, and strain through a paper filter.

Use: This works as a fluid for storing lenses.

Notes:

ALL-PURPOSE FACE AND HAND CREAM

Ingredients:			
Mineral Oil	14 oz.	392 g.
White Beeswax	1 oz.	28 g.
Paraffin Wax	3 oz.	84 g.
Petroleum Jelly	2 oz.	56 g.
Water	3 oz.	84 g.
Potash (Potassium Carbonate)	Spec	1 g.
Borax	Spec	1 g.
Perfume (Oil-Soluble)	to suit	

Mixing: Pour mineral oil in the top section of a double boiler. Add white beeswax, paraffin wax, and petroleum jelly to mineral oil in double boiler. Heat at a low temperature (about 150° F.) until all ingredients have melted together. Heat water in separate container and add potash and borax to it. Slowly pour this mixture into molten mixture in double boiler, with constant stirring, until an emulsion forms. Add oil-soluble perfume as desired. Pour into jars and allow to cool.

Use: Apply to face and hands as moisturizer.

CAUTIONS: Potash is toxic if taken internally.

Notes:

ANTIBACTERIAL SKIN CREAM

Ingredients:			
Powdered Paraffin Wax	3/4 C.	170 g.
Petrolatum	1/4 C.	57 g.
Cetyl Alcohol	1/2 C.	113 g.
Cetyltrimethylammonium Bromide	1 t.	4 g.
Water	3-1/2 C.	829 ml.

Mixing: In the top of a double boiler, melt powdered paraffin wax and petrolatum and stir together. Then stir in cetyl alcohol and cetyltrimethylammonium bromide. Add water slowly, stirring constantly. Cool to just above solidification point, pour into jars, and cool to room temperature.

Use: Use as a first-aid ointment.

Notes:

ANTICHAP HAND CREAM I

Ingredients:			
Glycerin		2/3 C.	158 ml.
Denatured Alcohol or			
Isopropyl Alcohol		1/3 C.	79 ml.
Peanut Oil		2 T.	30 ml.
Tragacanth, Powdered		1 t.	4 g.
Tincture of Benzoin		2 t.	10 ml.

Mixing: Mix glycerin into alcohol. Stir in peanut oil, tragacanth, and tincture of benzoin, in that order.

Use: Pour small amount on hand and rub into skin, especially at knuckles, wrists, and backs of hands.

CAUTIONS: Denatured and isopropyl alcohol may be toxic by ingestion and are flammable.

Notes: For proper grade of alcohol, see Appendix 4, *Denatured Alcohol.*

ANTICHAP HAND CREAM II

Ingredients:			
Mineral Oil		1-1/2 C.	355 ml.
Diglycol Laurate		1 C.	237 ml.
Camphor Oil		1 T.	15 ml.
Water		to suit	
Water-base Perfume		to suit	

Mixing: Stir mineral oil, diglycol laurate, and camphor oil together. Then slowly add water, stirring to desired thickness. Then add desired amount of perfume.

Use: Apply to hands sparingly.

Notes:

ANTICHAP HAND CREAM III

Ingredients:			
Camphor		2-1/4 T.	31 g.
Glycerin		1/2 C.	118 ml.
Water-base Perfume		to suit	
Denatured Alcohol or			
Isopropyl Alcohol		1/2 C.	118 ml.

Mixing: Stir camphor, glycerin, and perfume into denatured or isopropyl alcohol.

Use: Pour a small amount into the palm of your hand and rub in well.

CAUTIONS: Camphor vapors are flammable. Denatured and isopropyl alcohol may be toxic by ingestion and are flammable.

Notes: For proper grade of alcohol, see Appendix 4, *Denatured Alcohol.*

CLEANSING CREAM I

Ingredients:			
Mineral Oil	2-1/2 C.	592 ml.
Paraffin	1-1/4 C.	284 g.
Petrolatum	1 C.	227 g.
Oil-base Perfume	to suit	

Mixing: Melt mineral oil, paraffin, and petrolatum in the top of a double boiler. Turn off heat and stir in perfume to suit. Cool to just above solidification point, about 160° F., and pour into containers. Cool to room temperature before covering.

Use: Use as a cold cream, to remove makeup, and to clean skin.

Notes:

CLEANSING CREAM II

Ingredients:			
Ammonium Stearate Paste	25 oz.	700 g.
White Mineral Oil	2-1/2 oz.	70 g.
Oil-soluble Perfume	to suit	

Mixing: Stir ammonium stearate paste and mineral oil together vigorously until ammonia smell disappears, then add perfume and transfer to jars.

Use: Use as a cleansing cream and to remove cosmetics.

Notes:

COLD CREAM

Ingredients:			
	Mineral Oil	2 C.	474 ml.
	White Beeswax	1/2 C.	113 g.
	Water	1 C.	237 ml.
	Borax	1-1/2 T.	21 g.

Mixing: Heat mineral oil and white beeswax in the top of a double boiler until the wax melts and can be mixed with the oil. Cool down to 120° F. Separately heat water to 120° and stir in borax. Pour the mixture of borax and water into the mixture of mineral oil and white beeswax slowly, stirring constantly. When it just begins to solidify, pour into jars and cool to room temperature.

Use: Use as a cleansing cream and to remove cosmetics.

Notes:

DEPILATORY CREAM

Ingredients:			
	Barium Sulfide	3 T.	42 g.
	White Petrolatum	1-1/4 C.	284 g.
	Spermaceti	6 T.	84 g.
	Stearic Acid	6 T.	84 g.
	Potash (Potassium Carbonate)	1 T.	14 g.
	Water	1-3/4 C.	414 ml.

Mixing: Mix all ingredients together with a fork to form a creamy consistency.

Use: Apply with a cotton swab, leave on for five minutes, and wash off.

CAUTIONS: Barium sulfide and potash are both toxic if taken internally.

Notes:

GERMICIDAL CREAM

Ingredients:			
	White Petrolatum	1/2 C.	113 g.
	Mineral Oil	6 T.	90 ml.
	Beeswax	2 T.	28 g.
	Water	1/4 C.	59 ml.
	Borax	1 T.	14 g.
	Parachlorometacresol	1 T.	15 ml.

Mixing: Heat petrolatum, mineral oil, and beeswax in the top of a double boiler until the beeswax melts. Separately heat water and borax, stirring gently until borax is dissolved. Do not boil. Turn off all heat and mix the two solutions together. Finally, add parachlorometacresol. Cool until the mixture just starts to solidify, pour into jars, and cool to room temperature.

Use: Use as a first-aid cream.

Notes:

HONEY AND ALMOND CREAM

Ingredients:			
	Stearic Acid	3/4 T.	10 g.
	Ethylene Glycol	1 T.	15 ml.
	Glycerin	3 T.	45 ml.
	Honey	1 t.	5 ml.
	Water	1-1/4 C.	296 ml.
	Almond Oil	to suit	

Mixing: Heat stearic acid, ethylene glycol, and glycerin in the top of a double boiler to 150° F. Separately heat honey and water to 150° F. Combine the two mixtures and allow to cool for a few minutes. Then stir in almond oil. Pour into containers and cool to room temperature.

Use: Rub on lightly with finger tips.

CAUTIONS: Almond oil vapors are toxic.

Notes:

LANOLIN HAND AND FACE CREAM

Ingredients:			
	Powdered Soap	1/2 t.	2 g.
	Distilled Water	2 T.	30 ml.
	Lanolin	2 T.	28 g.
	Glycerin	2 T.	30 ml.

Mixing: Dissolve powdered soap in distilled water, then stir in lanolin and glycerin and warm just until all can be easily mixed together. Pour into containers.

Use: Use as a smoothing and moisturizing skin cream.

Notes:

LIME AND GLYCERIN CREAM

Ingredients:			
Glycerin	1/8 t.	0.6 ml.
Lime Water	1 C.	237.0 ml.
Almond Oil	1/8 t.	0.6 ml.
Lemon Oil	1/8 t.	0.6 ml.

Mixing: Mix glycerin and lime water, then stir in almond oil and lemon oil. Store in jar or plastic squeeze bottle.

Use: Pour a small amount in palm and rub lightly on face and neck.

CAUTIONS: Almond oil vapors are toxic.

Notes:

"NIVEA"-TYPE CREAM

Ingredients:			
Anhydrous Lanolin	3/4 T.	10 g.
White Petrolatum	1 C.	227 g.
Water	2 C.	474 ml.
Glycerin	1 t.	5 ml.
Water-base Perfume	to suit	

Mixing: Heat anhydrous lanolin and petrolatum in the top of a double boiler to about 150° F., or until liquified. Separately heat water and glycerin to about the same temperature and stir the two mixtures together. Cool until it just starts to solidify, then stir in perfume, and pour into jars. Cool to room temperature before covering.

Use: Use as a skin softener.

Notes:

NOURISHING CREAM

Ingredients:			
Anhydrous Lanolin	2 T.	28.0 g.
Stearic Acid	3/4 T.	10.0 g.
Triethanolamine	1/2 t.	2.5 ml.
Water	1 C.	237.0 ml.
Water-base Perfume	to suit	

Mixing: In the top of a double boiler heat anhydrous lanolin, stearic acid, and triethanolamine until melted. Stir together, turn off heat, and add water and perfume. Stir thoroughly, cool to room temperature, and bottle.

Use: Rub into hands and face, elbows, and other exposed areas of skin in the evening. Leave on overnight. Wipe or wash off in the morning.

CAUTIONS: Triethanolamine may be somewhat irritating to skin and mucous membranes.

Notes:

SUN CREAM

Ingredients:			
Zinc Oxide	1/2 C.	113 g.
Lanolin	1 C.	227 g.
Water	1/2 C.	118 ml.
Water-base Perfume	to suit	

Mixing: Mix zinc oxide and lanolin into a paste. Slowly add water, mixing constantly, then stir in perfume. Store in container.

Use: Coat vulnerable areas, such as nose and forehead, before going in the sun. In severe sun, use sun screen.

CAUTIONS: Zinc oxide is poisonous if taken internally.

Notes:

DANDRUFF TREATMENT

Ingredients:			
Ammonium Carbonate	2 T.	28.0 g.
Isopropyl Alcohol	3/4 C.	177.0 ml.
Glycerin	1/2 C.	118.0 ml.
Water	5 C.	1.2 L.
Water-base Perfume	to suit	

Mixing: Stir ammonium chloride, isopropyl alcohol, glycerin, and perfume into water. Store in a jar or plastic container.

Use: For control of dandruff, rub a small amount into the scalp between shampoos.

CAUTION: Ammonium carbonate, if heated, may result in irritating fumes. Isopropyl alcohol may be toxic by ingestion and is flammable.

Notes: For discussion on grades of alcohol, see Appendix 4, *Denatured Alcohol.*

DENTAL TOOTH POWDER

Ingredients:			
Sodium Chloride (Salt)	3/4 C.	170.0 g.
Sodium Bicarbonate (Baking Soda)	1/4 C.	57.0 g.
Magnesium Carbonate	2 T.	28.0 g.
Sodium Perborate	1 T.	14.0 g.
Peppermint Oil	15 Drops	1.5 ml.

Mixing: Mix the dry ingredients first, then stir in the peppermint oil.

Use: Pick-up amount to cover a wet toothbrush.

Notes:

AMMONIATED DENTRIFICE POWDER

Ingredients:		
Dibasic Ammonium Phosphate	3 t.	12 g.
Calcium Carbonate (Whiting)	1 C.	227 g.

Mixing: Combine ingredients in a bowl using a fork.

Use: Use on a wet toothbrush.

Notes:

DENTURE ADHESIVE

Ingredients:		
Boric Acid	2 T.	28 g.
Tragacanth, Powdered	3/4 C.	170 g.
Acacia, Powdered	1/2 C.	113 g.

Mixing: Stir ingredients together with a fork.

Use: Sprinkle lightly on wet dental plate.

Notes:

"You're using my athlete's foot ointment."

DENTURE CLEANER I

Ingredients: Citric Acid — 1 t. — 4.0 g.
Denatured Alcohol or
   Isopropyl Alcohol — 1 C. — 237.0 ml
Peppermint Oil — 10 Drops — 1.5 ml.

Mixing: Mix citric acid into denatured or isopropyl alcohol and add peppermint oil.

Use: Soak dentures at least 15 minutes; rinse in clean water.

CAUTIONS: Denatured and isopropyl alcohol may be toxic by ingestion and both are flammable.

Notes: For proper grade of alcohol, see Appendix 4, *Denatured Alcohol.*

DENTURE CLEANER II

Ingredients: Sodium Perborate — 1 T. — 14 g.
Hot Water — 1 C. — 237 ml.

Mixing: Dissolve sodium perborate into hot water.

Use: Immerse dentures in mixture after mixing. They are cleaned by the foaming action.

Notes:

DENTURE CLEANER III

Ingredients: Pumice Powder — 1/2 C. — 113 g.
Glycerin — 1/4 C. — 59 ml.

Mixing: Add pumice powder to glycerin, stirring to form a paste. Store in an airtight container.

Use: Brush dentures using a toothbrush, with this slightly abrasive paste.

Notes:

DEODORANT CREAM

Ingredients:			
Stearic Acid		3/4 C.	170 g.
Water		3-1/2 C.	829 ml.
Triethanolamine		2 T.	30 ml.
Aluminum Potassium Sulfate (Powdered Alum)		2 T.	28 g.

Mixing: Melt stearic acid in 2 cups (474 ml.) water. Turn off heat. Mix triethanolamine into the remaining water, and add this second mixture to the first. Cool and stir in aluminum potassium sulfate. As it just begins to thicken, pour into jars.

Use: Apply a small amount to underarms with your fingertips.

CAUTIONS: Triethanolamine may be somewhat irritating to skin and mucous membranes.

Notes:

DEODORANT LOTION FOR SKIN

Ingredients:			
Glycerin		1/2 oz.	15 ml.
Water		23 oz.	690 ml.
Boric Acid		1 oz.	28 g.
Salicylic Acid		1/4 oz.	7 g.
Water-Soluble Perfume		to suit	

Mixing: Mix glycerin into water with stirring. Add boric acid and salicylic acid. Continue to stir. As a final step, add perfume to suit.

Use: Apply to skin as desired. Discontinue if rash appears.

Notes:

DEODORANT PADS (UNDERARM)

Ingredients:			
Aluminum Potassium Sulfate (Alum)		3 T.	42 g.
Water		1 pt.	473 ml.
2 Inch-Diameter Flannel Discs			
Water-Soluble Perfume		to suit	

Mixing: Mix aluminum potassium sulfate into water. Add water-soluble perfume as desired. Place stack of pads into a jar about 2-1/4-inch in diameter. Pour mixture over pads to saturate.

Notes: An excellent place to buy flannel pads is a sporting-goods store where

the pads are economical and available as gun-cleaning swatches.
Notes:

DEODORANT POWDER

Ingredients: Aluminum Potassium Sulfate		
(Alum)	1/3 C.	76 g.
Boric Acid	1/4 C.	57 g.
Talc	1/3 C.	76 g.
Perfume (Oil-base)	to suit	

Mixing: Stir all ingredients together.

Use: Apply with shaker can or a plastic squeeze bottle.

Notes:

DEODORANT SPRAY (UNDERARM)

Ingredients: Aluminum Potassium Sulfate		
(Powdered Alum)	3 T.	42 g.
Water	1 pt.	473 ml.

Mixing: Mix aluminum potassium sulfate into water.

Use: Apply to underarms with a plastic spray bottle or cloth.

Notes:

DEPILATORY AFTER-USE LOTION

Ingredients: Acetic Acid	1 T.	15 ml.
Zinc Acetate	8 T.	112 g.
Rose-Oil	1 T.	15 ml.
Water	3 C.	711 ml.

Mixing: Mix acetic acid, zinc acetate, and rose oil into water.

Use: Apply as a soothing lotion after the depilation.

Notes:

DUSTING POWDER

Ingredients: Talc   15 oz.   450 g.

Ingredients:	Talc	15 oz.	450 g.
	Boric Acid	5 oz.	130 g.
	Oil-soluble Perfume	to suit	

Mixing: Mix talc and boric acid thoroughly and add oil-soluble perfume as desired.

Use: Use for dusting the body after a bath.

Notes:

EARWAX SOFTENER

Ingredients:	Sodium Bicarbonate (Baking Soda)	1/8 t.	0.5 g.
	Water	1/2 C.	118.0 ml.
	Glycerin	1/4 C.	59.0 ml.

Mixing: Mix sodium bicarbonate and water, then stir in glycerin. Store in a bottle with a sterile dropper.

Use: A few drops in the ear will soften wax.

Notes:

EAU DE COLOGNE

Ingredients:	Denatured Alcohol	1 C.	237.0 ml.
	Lemon Oil	17 Drops	1.7 ml.
	Bergamot Oil	7 Drops	0.7 ml.
	Neroli Oil	40 Drops	4.0 ml.
	Rosemary Oil	7 Drops	0.7 ml.

Mixing: Mix the remaining ingredients into the isopropyl alcohol, stirring constantly. Store in an old perfume bottle or jar.

Use: Use as a splash-on refreshener.

CAUTIONS: Denatured alcohol may be toxic by ingestion and is flammable.

Notes: For proper grade of alcohol, see Appendix 4, *Denatured Alcohol.*

ELECTRIC AFTER-SHAVE

Ingredients: Denatured Alcohol or		
Isopropyl Alcohol	1/4 C.	59 ml.
Glycerin	5 T.	75 ml.
Water	1 C.	237 ml.
Witch Hazel	1/4 C.	59 ml.
Aluminum Potassium Sulfate		
(Powdered Alum)	1/2 t.	2 g.
Boric Acid	1/4 t.	1 g.
Water-base Perfume	to suit	

Mixing: Mix denatured or isopropyl alcohol, glycerin, and water. Then add remaining ingredients, stirring well. Store in an airtight container.

Use: Use after shaving with an electric razor.

CAUTIONS: Denatured and isopropyl alcohol may be toxic by ingestion and both are flammable.

Notes: For proper grade of alcohol, see Appendix 4, *Denatured Alcohol.*

ELECTRIC PRESHAVE

Ingredients: Denatured Alcohol or		
Isopropyl Alcohol	1 C.	237 ml.
Glycerin	2 T.	30 ml.
Water	3/4 C.	177 ml.
Water-base Perfume	to suit	

Mixing: Stir all ingredients together. Store in an airtight container.

Use: Apply as a beard softener and skin lubricant before shaving with an electric razor.

CAUTIONS: Denatured and isopropyl alcohol may be toxic by ingestion and both are flammable.

Notes: For proper grade of alcohol, see Appendix 4, *Denatured Alcohol.*

EYE-SMOG PROTECTIVE COMPOUND

Ingredients: Sodium Bicarbonate (Baking Soda)   1/16 t.   0.3 g.
            Distilled Water   1 C.   237 ml.

Mixing: Dissolve sodium bicarbonate in water.

Use: Apply to eyes with eye cup or wash liberally with mixture.

Notes:

LENS CLEANER, EYEGLASS

Ingredients: Potassium Oleate   1/2 C.   113 g.
            Glycerin   1/4 C.   59 ml.
            Turpentine   10 Drops   1 ml.

Mixing: Melt potassium oleate and glycerin in the top of a double boiler, turn off heat, and stir in turpentine. Cool and pour into a jar.

Use: Apply with a soft, lint-free cloth. This is an excellent lens cleaner for eyeglasses, cameras, telescopes, and all fine lenses.

CAUTIONS: Turpentine is toxic if taken internally and flammable. Handle with care.

Notes:

EYELASH AND EYEBROW CONDITIONER AND DARKENER

Ingredients: Yellow Petroleum Jelly   1/4 C.   57 g.
            Castor Oil   1/4 C.   59 ml.
            Paraffin Wax   1/4 t.   1 g.

Mixing: Heat ingredients in the top of a double boiler until paraffin wax melts and can be mixed in. Cool and pour into jars.

Use: Apply small amount to eyelashes and eyebrows with a soft brush to condition them and also to darken them.

Notes:

EYELASH AND EYEBROW IMPROVER

Ingredients:			
Beeswax	1 C.	227 g.
Cocoa Butter	1-1/4 C.	284 g.
Peanut Oil	3 C.	711 ml.

Mixing: Melt beeswax and cocoa butter in the top of a double boiler, then add peanut oil. When partially cool, pour into jars.

Use: Apply a small amount to eyelashes and eyebrows with a soft brush.

Notes:

EYE WASH

Ingredients:			
Sodium Bicarbonate (Baking Soda)	1/8 t.	0.5 g.
Water	1 C.	237.0 ml.

Mixing: Dissolve sodium bicarbonate into water. Store in bottle with sterile dropper.

Use: Use with an eye-wash cup or an eyedropper. This provides excellent relief from smog and pollution.

Notes:

FACE LOTION I

Ingredients:			
Glycerin	1/4 C.	59 ml.
Isopropyl Alcohol	1-3/4 C.	414 ml.
Water	3 C.	711 ml.
Water-base Perfume	to suit	

Mixing: Stir ingredients together. Store in an airtight container.

Use: Apply as a refreshing face wash.

CAUTIONS: Isopropyl alcohol may be toxic by ingestion and is flammable.

Notes: For discussion on grades of alcohol, see Appendix 4, *Denatured Alcohol.*

FACE LOTION II

Ingredients:			
Glycerin	1/4 C.	59 ml.
Isopropyl Alcohol	1-3/4 C.	414 ml.
Witch Hazel	3 C.	711 ml.

Mixing: Stir ingredients together. Store in an airtight container.

Use: Apply as a refreshing face wash.

CAUTIONS: Isopropyl alcohol may be toxic if taken internally and is flammable.

Notes: For discussion on grades of alcohol, see Appendix 4, *Denatured Alcohol.*

FACE POWDER

Ingredients:			
Talc	4 C.	908 g.
Boric Acid	1/8 C.	28 g.
Cornstarch	1 C.	227 g.
Dye (Oil-soluble)	to suit color	
Perfume (Oil-soluble)	to suit	

Mixing: Mix talc, boric acid, and cornstarch in a bowl using a fork. Add dye to obtain desired color and perfume. Continue mixing. Store in plastic or glass container.

Use: Apply as a cosmetic face powder.

Notes:

FACE POWDER BASE I

Ingredients·			
Talc	4-1/2 C.	1 kg.
Precipitated Chalk	1 C.	227 g.
Zinc Oxide	1 C.	227 g.

Mixing: Mix ingredients and sift through a flour sifter. Store in a glass jar or plastic container.

Use: This is a basic white face powder. Color and/or perfume may be added as desired.

CAUTIONS: Zinc oxide is poisonous if taken internally.

Notes:

FACE POWDER BASE II

Ingredients:			
Talc		4 C.	908 g.
Boric Acid		2 T.	28 g.
Cornstarch		1 C.	227 g.

Mixing: Mix ingredients and sift through a flour sifter. Store in glass jar or plastic container.

Use: This is another basic white face powder. Oil-base perfume and dyes may be added to suit your color or odor preferences.

Notes:

FACE WATER REFRESHER

Ingredients:		
Denatured Alcohol	3/8 C.	90 ml.
Witch Hazel	3 T.	45 ml.
Glycerin	2 T.	30 ml.
Perfume (Water-soluble)	to suit	

Mixing: Into denatured alcohol, mix witch hazel and glycerin. Add perfume and stir. Store in an airtight container.

Use: Apply with cotton balls or flannel pads to refresh skin.

CAUTIONS: Denatured alcohol is toxic if taken internally and is flammable.

Notes: For proper grade of alcohol, see Appendix 4, *Denatured Alcohol.*

FACIAL BLEACH

Ingredients:		
Sodium Perborate	1-1/2 T.	21 g.
Calcium Carbonate (Chalk)	2 C.	454 g.

Mixing: Stir ingredients together with a fork. Store in glass jar or plastic container.

Use: Make a paste by adding a little water to the powder. Apply with a soft brush or cloth. Leave on overnight and wash off in the morning.

Notes:

FACIAL PORE CLOSER

Ingredients:			
Isopropyl Alcohol		1/2 C.	118 ml.
Water		1 C.	237 ml.
Aluminum Potassium Sulfate (Powdered Alum)		1 T	14 g.
Talc		1 C.	227 g.
Boric Acid		1 T.	14 g.

Mixing: Mix isopropyl alcohol and water. Add aluminum potassium sulfate, talc, and boric acid, stirring well. Store in an airtight container.

Use: Apply liberally with cotton swab, let dry, and rinse off.

CAUTIONS: Isopropyl alcohol is toxic by ingestion and is flammable.

Notes: For discussion on grades of alcohol, see Appendix 4, *Denatured Alcohol.*

FEMININE HYGIENE SPRAY DEODORANT

Ingredients:			
Aluminum Potassium Sulfate (Powdered Alum)		1 C.	227 g.
Boric Acid, Powdered		3-1/2 C.	795 g.

Mixing: Mix ingredients together with a fork. Store in a plastic squeeze bottle.

Use: Mix 1 tablespoon (14 g.) to 1 quart (1 L.) of warm water.

Notes:

FINGER STAIN REMOVER

Ingredients:			
Sodium Sulfate		2 T.	28 g.
Water		1/2 C.	118 ml.

Mixing: Mix sodium sulfate into water and stir until dissolved. Store in a plastic container.

Use: Rub on stained area with a cotton swab, let stand for 15 minutes, and wash off.

Notes:

FINGERNAIL BLEACH

Ingredients: Sodium Perborate   1/4 C.   57 g.

               Water   3/4 C.   177 ml.

Mixing: Mix sodium perborate into water until dissolved. Store in a plastic container.

Use: Soak nails in this solution for about 15 minutes. Repeat as necessary, but let dry between soakings.

Notes:

CUTICLE CREAM

Ingredients: Mineral Oil   1 C.   237 ml.

               Paraffin Wax   2 T.   28 g.

               Oil-base Perfume   to suit

Mixing: Heat mineral oil and paraffin wax in the top of a double boiler until paraffin wax is melted and can be mixed in. Cool and add an oil-base perfume. Pour into jars.

Use: Apply a small amount with cotton around nails on the cuticle.

Notes:

CUTICLE REMOVER I

Ingredients: Caustic Potash (Potassium Hydroxide)   1/2 t.   2 g.

               Water   2 C.   474 ml.

               Isopropyl Alcohol   1 T.   15 ml.

Mixing: Stir all the ingredients together. Pour into an airtight container.

Use: Apply with cotton swab, let stand for 2 or 3 minutes, work cuticles with a cuticle stick.

CAUTIONS: Caustic potash heats on contact with water and can cause severe burns to skin. Handle with care. Store in an airtight container. Isopropyl alcohol is toxic by ingestion and is flammable.

Notes: For discussion on grades of alcohol, see Appendix 4, *Denatured Alcohol.*

CUTICLE REMOVER II

Ingredients:			
Trisodium Phosphate	1 T.	14 g.
Glycerin	3 T.	45 ml.
Water	to suit	

Mixing: Mix trisodium phosphate and glycerin, stirring with a spoon until the trisodium phosphate is dissolved. Then add water until the preferred consistency is reached. Store in an airtight container.

Use: Apply with cotton swab, let stand 2 or 3 minutes, work cuticles with a cuticle stick.

CAUTIONS: Trisodium phosphate is a skin irritant and moderately toxic by ingestion. Use rubber gloves while handling chemical.

Notes:

FINGERNAIL HARDENER

Ingredients:			
Glycerin	1 T.	15 ml.
Water	3 T.	45 ml.
Aluminum Potassium Sulfate (Powdered Alum)	1 t.	4 g.

Mixing: Mix glycerin and water, then add aluminum potassium sulfate, stirring until dissolved. Store in an airtight container.

Use: Coat nails at night and remove in the morning with denatured alcohol.

CAUTIONS: Denatured alcohol is toxic by ingestion and is flammable.

Notes: For proper grade of denatured alcohol, see Appendix 4, *Denatured Alcohol.*

NAIL-POLISHING POWDER

Ingredients:			
Pumice Powder	4 T.	56 g.
Talc	1 T.	14 g.
Stannic Oxide	5 T.	70 g.

Mixing: Stir all ingredients together. Store in an airtight container.

Use: Apply to nails with a damp cloth and polish. Rinse in clean water.

Notes:

FINGERNAIL SOFTENER

Ingredients:			
Triethanolamine	3 T.	45 ml.
Water	1/2 C.	118 ml.
Castor or Olive Oil	2 T.	30 ml.

Mixing: Mix triethanolamine into water, and slowly stir in castor or olive oil.

Use: Paint on nails at night, remove in the morning with water.

CAUTIONS: Triethanolamine may be somewhat irritating to skin and mucous membranes.

Notes:

FINGERNAIL STAIN REMOVER

Ingredients:			
Tartaric Acid (Cream of Tartar)	1 T.	14 g.
Water	3 T.	45 ml.

Mixing: Dissolve tartaric acid in water.

Use: Apply to stains with a cloth or tissue, leave on for 10 minutes, and rinse off.

Notes:

NAIL WAX–CLEAR LIQUID

Ingredients:			
Petrolatum	1 t.	4 g.
Paraffin Wax	1 T.	14 g.
Ethyl Acetate	1 T.	15 ml.
Deodorized Kerosene	1 C.	237 ml.

Mixing: Heat ingredients together in the top of a double boiler, using only enough heat to form a homogenous mass. Store in an airtight container.

Use: Apply to nails as desired.

CAUTIONS: Ethyl acetate is moderately toxic by inhalation and skin absorption, irritating to eyes and skin, and highly combustible. Deodorized kerosene is toxic if taken internally and flammable.

Notes:

FOOT BATH

Ingredients: Menthol Crystals	1/8 t.	0.5 g.
Aluminum Potassium Sulfate		
(Powdered Alum)	4 T.	56.0 g.
Boric Acid	1/2 C.	113.0 g.
Magnesium Sulfate	2/3 C.	151.0 g.

Mixing: Stir ingredients together with a fork.

Use: Use 1 teaspoon (4 g.) per gallon (4 L.) of hot water.

Notes:

FOOT PERSPIRATION LOTION

Ingredients: Hexamethylenetetramine	1/2 t.	2 g.
Tragacanth	1/2 t.	2 g.
Water	3 C.	711 ml.
Talc	1 C.	227 g.

Mixing: Mix hexamethylenetetramine and tragacanth into water. Slowly stir in talc. Store in bottles.

Use: Wipe on feet with cotton swab. Allow to dry before putting on socks.

CAUTIONS: Hexamethylenetetramine is moderately toxic and flammable.

Notes:

FOOT POWDER

Ingredients: Cornstarch	1-3/4 C.	397 g.
Boric Acid	1/4 C.	57 g.

Mixing: Mix ingredients with a fork. Store in plastic squeeze bottle.

Use: Sprinkle in shoes and directly on feet.

Notes:

OXYGEN FOOT POWDER

Ingredients:			
Sodium Thiosulfate	1 T.	14 g.
Sodium Perborate	1 T.	14 g.
Sodium Borate	6 T.	84 g.
Sodium Bicarbonate (Baking Soda)	3/4 C.	170 g.

Mixing: Stir ingredients together with a fork. Store in jar or an airtight container.

Use: Use 2 tablespoons (28 g.) in 2 quarts (2L.) of hot water for a foot bath. The released oxygen has a very soothing effect.

Notes:

BREATH FRESHENER

Ingredients:			
Chloramine (5 gr. tablet)	1 tablet	5 gr.
Water	1 oz.	30 ml.

Mixing: Mix ingredients thoroughly. Store in a jar or an airtight container.

Use: This formula may be used on a toothbrush or as a mouthwash.

Note:

GLYCERIN SKIN GEL

Ingredients:			
Water	2-1/4 C.	533 ml.
Gelatin	5 t.	20 g.
Glycerin	3 T.	45 ml.
Water-base Perfume	to suit	

Mixing: Heat 1 cup (237 ml.) of water to dissolve gelatin. To the balance of water, add glycerin, then mix the two solutions together before adding perfume, if desired.

Use: Use as a soothing, smoothing skin gel. Apply with a cotton swab or soft tissue.

Notes:

HAIR CLEANER

Ingredients:			
Ammonium Stearate	1-1/2 C.	340 g.
Water	1/2 C.	118 ml.
Water-base Perfume	to suit	

Mixing: Into ammonium stearate, mix water; add water-base perfume. You may vary the amount of water to obtain a consistency you like.

Use: Use as a paste-type shampoo.

Notes:

HAIR CONDITIONER I

Ingredients:			
Stearamide	2 T.	28 g.
Soap Chips	2 T.	28 g.
Mineral Oil	1/2 C.	118 ml.
Beeswax	1 t.	4 g.
Water	1 C.	237 ml.

Mixing: Heat all ingredients together in the top of a double boiler just until they can be stirred together. When almost cool, pour into jars.

Use: Apply a small amount as a hair dressing.

Notes:

"Oh, no, split-ends!"

HAIR CONDITIONER II

Ingredients:			
Stearic Acid	2 t.	8 g.
Petrolatum	1/4 C.	57 g.
Lanolin	1 t.	4 g.
Triethanolamine	1 t.	4 g.
Water	1/2 C.	118 ml.
Water-base Perfume	to suit	

Mixing: Melt stearic acid, petrolatum and lanolin in the top of a double boiler. Heat triethanolamine and water over medium-heat. Remove both from heat. Add the two mixtures together, and when cool, add a water-base perfume. Store in an airtight container.

Use: Apply a small amount as a hair dressing.

CAUTIONS: Triethanolamine may be somewhat irritating to skin and mucous membranes.

Notes:

HAIR CONDITIONER III

Ingredients:			
Paraffin Wax	1 t.	4.0 g.
Paraffin Oil	1/4 C.	59.0 ml.
Borax	1/8 t.	0.5 g.
Water	1/2 T.	7.5 ml.
Oil-base Perfume	to suit	

Mixing: Melt paraffin wax in 1/8 cup (30 ml.) of paraffin oil in the top of a double boiler. Stirring vigorously, add balance of paraffin oil. Dissolve borax in water and add to first mixture. While cooling, stir into creamy consistency. Then add an oil-base perfume to suit. Store in an airtight container.

Use: Apply small amount as a hair dressing.

Notes:

HAIR CREAM

Ingredients:			
Paraffin Wax		1 t.	4.0 g.
Paraffin Oil		1/4 C.	59.0 ml.
Borax		1/8 t.	0.5 g.
Distilled Water		1/2 T.	7.5 ml.
Water-soluble Perfume			

Mixing: Melt paraffin wax in 1/8 cup (30 ml.) of paraffin oil in the top of a double boiler. Add balance of paraffin oil with vigorous stirring. Dissolve borax and distilled water and add to first mixture. Stir into a creamy consistency and cool. Add a water-soluble perfume to suit. Store in an airtight container.

Use: Apply sparingly to manage hair.

Notes:

GREASELESS HAIR DRESSING FOR MEN

Ingredients:			
Glycerin		1 C.	237 ml.
Isopropyl Alcohol		1 C.	237 ml.
Water-base Perfume		to suit	

Mixing: Stir ingredients together thoroughly. Store in a plastic jar or can.

Use: Use a small amount to help control hair.

CAUTIONS: Isopropyl alcohol may be toxic by ingestion and is flammable.

Notes: For discussion on grades of alcohol, see Appendix 4, *Denatured Alcohol.*

HAIR OIL

Ingredients:			
Isopropyl Alcohol		1/2 C.	118 ml.
Glycerin		1/4 C.	59 ml.
Water		6 T.	90 ml.
Water-base Perfume		to suit	

Mixing: Stir all ingredients together thoroughly. Store in a plastic jar or can.

Use: Use a small amount on hair as a managing oil.

CAUTIONS: Isopropyl alcohol may be toxic by ingestion and is flammable.

Notes: For discussion on grades of alcohol, see Appendix 4, *Denatured Alcohol.*

WIG AND HAIR-PIECE CLEANING POWDER

Ingredients:			
Talc	1 C.	227 g.
Zinc Oxide	1 T.	14 g.
Calcium Carbonate (Chalk)	2 T.	28 g.
Boric Acid	1 t.	4 g.

Mixing: Stir ingredients together with a fork. Store in a plastic jar, can, or squeeze bottle.

Use: Dust on and brush out thoroughly.

CAUTIONS: Zinc oxide is poisonous if taken internally.

Notes:

BLOND HAIR RINSE

Ingredients:			
Tincture of Rhubarb	3 T.	45 ml.
Isopropyl Alcohol	1/3 C.	79 ml.
Water	3-3/4 C.	888 ml.
Propylene Glycol	2 t.	10 ml.
Water-Base Perfume	to suit	

Mixing: Mix tincture of rhubarb and isopropyl alcohol into water, then add propylene glycol and a water-base perfume. Store in an airtight container.

Use: Use as a rinse after shampooing.

CAUTIONS: Isopropyl alcohol is toxic by ingestion and is flammable.

Notes: For discussion on grades of alcohol, see Appendix 4, *Denatured Alcohol.*

LEMON HAIR RINSE

Ingredients:			
Lemon Extract	10 Drops	3 ml.
Isopropyl Alcohol	1 C.	237 ml.
Citric Acid	2 T.	28 g.
Tartaric Acid	1/4 C.	57 g.
Water	1-1/2 C.	355 ml.

Mixing: Dissolve lemon extract in isopropyl alcohol. Separately mix the remaining ingredients. Finally, add the two mixtures together and bottle.

Use: Use as a rinse after shampooing.

CAUTIONS: Isopropyl alcohol is toxic by ingestion and is flammable.

Notes: For discussion on grades of alcohol, see Appendix 4, *Denatured Alcohol.*

LIQUID HAIR SET·

Ingredients:			
Acacia	1/2 t.	2 g.
Borax	1 t.	4 g.
Warm Water	2 C.	474 ml.
Isopropyl Alcohol	6 T.	90 ml.
Water-base Perfume	to suit	

Mixing: Dissolve acacia and borax in water. Then stir in isopropyl alcohol and add a water-base perfume. This formula may be diluted with water to a sprayable consistency if desired. Store in an open-mouth jar or plastic squeeze bottle.

Use: Dip comb in solution, or dilute and spray on.

CAUTIONS: Isopropyl alcohol is toxic by ingestion and is flammable.

Notes: For discussion on grades of alcohol, see Appendix 4, *Denatured Alcohol.*

HAIR SHAMPOO

Ingredients:			
Oleic Acid	1-1/4 C.	296 ml.
Coconut Oil	1 C.	237 ml.
Triethanolamine	1-1/4 C.	296 ml.
Water-base Perfume	to suit	

Mixing: Stir ingredients together thoroughly.

Use: This is a concentrate. Add enough water to reach the consistency you prefer.

CAUTIONS: Triethanolamine may be somewhat irritating to skin and mucous membranes.

Notes:

HAIR STRAIGHTENER

Ingredients:			
Paraffin Wax	3 oz.	84 g.
Petrolatum	7 oz.	196 g.
Oil-soluble Perfume	to suit	

Mixing: Place paraffin wax and petrolatum in the top of a double boiler and melt together while stirring. When cool add an oil-soluble perfume and transfer to jars.

Use: This may be applied to the hair with a brush or cloth.

Notes:

HAIR TONIC I

Ingredients:
Deodorized Castor Oil	1-1/2 T.	22 ml.
Isopropyl Alcohol	1 C.	237 ml.
Oil-base Perfume	to suit	

Mixing: Mix all ingredients together. Store in a jar or plastic container.

Use: Rub a small amount into the hair and scalp.

CAUTIONS: Isopropyl alcohol is toxic by ingestion and is flammable.

Notes: For discussion on grades of alcohol, see Appendix 4, *Denatured Alcohol.*

HAIR TONIC II

Ingredients:
Oxyquinoline Sulfate	1 t.	4 g.
Denatured Alcohol or Isopropyl Alcohol	4 T.	60 ml.
Water-base Perfume	to suit	
Distilled Water	3 C.	711 ml.

Mixing: Mix oxyquinoline sulfate, denatured or isopropyl alcohol, and a water-base perfume into distilled water. Store in a jar or plastic container.

Use: Rub a small amount into the hair and scalp.

CAUTIONS: Oxyquinoline sulfate is moderately toxic in concentrated form. Denatured and isopropyl alcohol may be toxic by ingestion and are flammable.

Notes: For proper grade of alcohol, see Appendix 4, *Denatured Alcohol.*

HAIR WAVE LOTION

Ingredients:
Tragacanth Gum	1/2 t.	2 g.
Isopropyl Alcohol	1/2 C.	118 ml.
Water	1-1/2 C.	355 ml.
Potash (Potassium Carbonate)	2 t.	8 g.
Borax	1/2 t.	2 g.
Water-base Perfume	to suit	

Mixing: Dissolve tragacanth gum in isopropyl alcohol and stir in water. Then add remaining ingredients. Store in jar or plastic container.

Use: Dip a comb in this solution to aid in styling.

CAUTIONS: Isopropyl alcohol may be toxic by ingestion and is flammable.

Notes: For discussion on grades of alcohol, see Appendix 4, *Denatured Alcohol.*

PERMANENT WAVE SOLUTION

Ingredients:			
Glycerin		1 oz.	30.0 ml.
Water		1-1/2 oz.	45.0 ml.
Isopropyl Alcohol		6 oz.	180.0 ml.
Tincture of Benzoin		10 oz.	300.0 ml.
Methyl Cellulose		2-1/2 oz.	70.0 g.
Venice Turpentine		1/4 oz.	7.5 ml.
Water-soluble Perfume		to suit	

Mixing: Mix glycerin and water into isopropyl alcohol and tincture of benzoin and stir. Add methyl cellulose and then venice turpentine. Finally, add perfume and store in jar or plastic squeeze bottle.

Use: Apply to waves in hair to make them permanent; rinse with clear-water (Note: time required depends upon hair type; trials needed).

CAUTIONS: Isopropyl alcohol and venice turpentine are flammable and moderately toxic.

Notes: For discussion on grades of alcohol, see Appendix 4, *Denatured Alcohol.*

DRY HAND CLEANER

Ingredients:			
Soap Powder		1 C.	227 g.
Fine Sawdust		1 C.	227 g.
Borax Powder		1 T.	14 g.

Mixing: Stir ingredients together. Store in plastic jar or can.

Use: Wet hands and shake off. Rub about 1 teaspoon (4 g.) of the powder into hands and rinse.

Notes:

HAND CLEANER FOR SOOT AND GREASE

Ingredients:	Diglycol Stearate	2 T.	30 ml.
	Mineral Oil	2/3 C.	158 ml.

Mixing: Stir ingredients together. Store in plastic jar or can.

Use: Rub a small amount in well and rinse with clear water.

Notes: This solution may be thinned with water if desired.

HAND LOTION

Ingredients:	Glycerin	6 T.	90 ml.
	Anhydrous Lanolin	6 T.	84 g.
	Petrolatum	1/2 C.	113 g.
	Boric Acid	1 T.	14 g.
	Oil-base Perfume	to suit	

Mixing: Heat glycerin, anhydrous lanolin, and petrolatum in a double boiler until all ingredients are liquified, then add boric acid and an oil-base perfume. Stir well, cool, and bottle.

Use: Apply as a soothing lotion for dry hands.

Notes:

HAND LOTION FOR CHAPPED HANDS

Ingredients:	Camphor	2-1/4 T.	32 g.
	Glycerin	1/2 C.	118 ml.
	Water-base Perfume	to suit	
	Denatured Alcohol	1/2 C.	118 ml.

Mixing: Mix ingredients into denatured alcohol and bottle.

Use: Apply as a soothing lotion for dry hands.

CAUTIONS: Camphor vapors are flammable. Denatured alcohol may be toxic by ingestion and is flammable.

Notes: For proper grade of alcohol, see Appendix 4, *Denatured Alcohol.*

COOLING HAND LOTION

Ingredients:		
Menthol	1/8 t.	0.5 g.
Denatured Alcohol or		
Isopropyl Alcohol	1-1/2 C.	355.0 ml.
Glycerin	1/4 C.	59.0 ml.
Water	3/4 C.	177.0 ml.
Water-base Perfume	to suit	

Mixing: Dissolve menthol in denatured or isopropyl alcohol, then stir in glycerin and water. Add a water-base perfume, if desired, and bottle.

Use: Apply as a soothing hand lotion for dry hands.

CAUTIONS: Denatured and isopropyl alcohol may be toxic by ingestion and are flammable.

Notes: For proper grade of alcohol, see Appendix 4, *Denatured Alcohol.*

PROTECTIVE HAND CREAM I

Ingredients:		
Talc	1/2 C.	113 g.
Zinc Oxide	1/2 C.	113 g.
Petrolatum, White	1 C.	227 g.

Mixing: Mix talc and zinc oxide into petrolatum; transfer to jars.

Use: Apply a light coat to hands before exposure to any hard work around the house or yard.

CAUTIONS: Zinc oxide is poisonous if taken internally.

Notes:

PROTECTIVE HAND CREAM II

Ingredients:		
Water	1-1/2 C.	355 ml.
Gelatin	1 T.	14 g.
Glycerin	3/4 C.	177 ml.

Mixing: Heat glycerin and water, stir in gelatin until dissolved. Cool and bottle.

Use: Use a thin film on hands before working.

Notes:

MECHANIC'S HAND SOAP

Ingredients: Denatured Alcohol or

Isopropyl Alcohol	1-3/4 C.	414 ml.
Stoddard Solvent	1/3 C.	79 ml.
Coconut Oil	2 C.	474 ml.
Lemon Oil	to suit	

Mixing: Mix denatured or isopropyl alcohol and stoddard solvent. Slowly stir in coconut oil, then add lemon oil to suit.

Use: Pour about 1 teaspoon (5 ml.) on hands. Rub well and rinse.

CAUTIONS: Denatured and isopropyl alcohol may be toxic if taken internally and are flammable. Stoddard solvent is mildly flammable.

Notes: For proper grade of alcohol, see Appendix 4, *Denatured Alcohol.*

WATERLESS HAND SOAP

Ingredients:		
Water	5 C.	1.2 L.
Powdered Soap	1/2 C.	113.0 g.
Potash (Potassium Carbonate)	2 T.	28.0 g.
Trisodium Phosphate	1/2 T.	7.0 g.
Asbestos Powder	1/2 C.	113.0 g.
Lemon Extract	to suit	

Mixing: Heat water, but do not boil. Stir in the remaining ingredients, cool, and pour into containers.

Use: Rub a small amount into hands and wipe off with a paper towel.

CAUTIONS: Potash is toxic if taken internally. Trisodium phosphate is a skin irritant; use rubber gloves. It is also moderately toxic by ingestion. **Do not inhale dust of asbestos powder.**

Notes:

FACE LOTION ASTRINGENT

Ingredients:			
Water		10 oz.	300 ml.
Isopropyl Alcohol		6 oz.	180 ml.
Talc		3-1/2 oz.	98 g.
Boric Acid		2 oz.	56 g.
Aluminum Potassium Sulfate (Powdered Alum)		2-1/2 oz.	70 g.
Water-soluble Perfume		to suit	

Mixing: Mix water and isopropyl alcohol and then mix in talc while stirring. Add boric acid, aluminum potassium sulfate, and a water-soluble perfume and continue stirring.

Use: Apply as an astringent face lotion to assist in tightening wrinkles.

CAUTIONS: Isopropyl alcohol may be moderately toxic by ingestion and is flammable.

Notes: For discussion on grades of alcohol, see Appendix 4, *Denatured Alcohol.*

LEG LOTION

Ingredients:			
Green Lettuce		1 C.	227 g.
Water		1 C.	237 ml.
Water-soluble Perfume or Cologne		to suit	

Mixing: Heat green lettuce and water while stirring. Allow to simmer for 1 hour. Strain off liquid and add a water-soluble perfume as desired.

Use: Apply to legs for a soothing lotion.

Notes:

LINIMENT–ATHLETIC

Ingredients:			
Olive Oil		1/2 C.	118 ml.
Isopropyl Alcohol		4 C.	948 ml.
Oil of Mustard		1 t.	5 ml.
Camphor		1 speck	1 g.
Menthol		1 speck	1 g.

Mixing: Thoroughly mix olive oil into isopropyl alcohol while stirring and then add oil of mustard, camphor, and menthol and continue stirring. Bottle.

Use: Apply to sore muscles and joints.

CAUTIONS: Care should be used, as this is a strong liniment which will heat. Do not allow to remain in contact with skin where undue redness occurs. Camphor vapors are flammable. Isopropyl alcohol may be toxic by ingestion and is flammable.

Notes: For discussion on grades of alcohol, see Appendix 4, *Denatured Alcohol.*

LINIMENT FOR GENERAL USE

Ingredients:		
Methyl Salicylate	6 T.	90 ml.
Olive Oil	12 T.	180 ml.

Mixing: Mix methyl salicylate into olive oil while stirring. An oil-soluble perfume may be added if desired.

Use: Apply as any rubbing liniment would be used.

CAUTIONS: Methyl salicylate is highly toxic by ingestion in concentrated form.

Notes:

RHEUMATIC AND ARTHRITIC LINIMENT

Ingredients:		
Menthol Crystals	1/4 C.	57 g.
Camphor Oil	6 T.	90 ml.
Olive Oil	5 T.	75 ml.
Methyl Salicylate	1-1/4 C.	296 ml.
Denatured Alcohol or		
Isopropyl Alcohol	4 T.	60 ml.

Mixing: Dissolve menthol crystals in camphor oil, then stir in olive oil, methyl salicylate, and denatured or isopropyl alcohol. Store in an airtight container.

Use: Apply as a liniment for rheumatism and arthritis.

CAUTIONS: Methyl salicylate is highly toxic by ingestion in concentrated form. Denatured and isopropyl alcohol may be toxic by ingestion and are flammable.

Notes: For discussion on grades of alcohol, see Appendix 4, *Denatured Alcohol.*

LIPSTICK

Ingredients:			
	Beeswax	3 T.	42 g.
	Castor Oil	1 T.	15 ml.
	Ceresin Wax	6 T.	84 g.
	Anhydrous Lanolin	1 T.	14 g.
	White Petroleum Jelly	6 T.	84 g.
	Oil-base Dye	to suit	

Mixing: Heat all ingredients in the top of a double boiler, stirring until well mixed. Partially cool and pour into small molds.

Use: This may be colored to use as regular lipstick, or uncolored to use as lip protective coating.

Notes:

LIPSTICK REMOVER

Ingredients:			
	Beeswax	1/2 C.	113 g.
	Stearic Acid	1/4 C.	57 g.
	Mineral Oil	1/2 C.	118 ml.

Mixing: Melt and mix ingredients in the top of a double boiler. Cool and pour into jars.

Use: Apply with tissue to remove lipstick and other makeup.

Notes:

MAKEUP REMOVER I

Ingredients:			
	Beeswax	1/2 C.	113 g.
	Paraffin Wax	1/3 C.	76 g.
	Mineral Oil	1 C.	237 ml.

Mixing: Melt beeswax and paraffin wax in the top of a double boiler. Stir in mineral oil. Cool and pour in jars.

Use: Apply with tissue and wipe off.

Notes:

MAKEUP REMOVER II

Ingredients:
White Petrolatum	1/2 C.	113 g.
Mineral Oil	3/4 C.	177 ml.
Isopropyl Alcohol	2 T.	30 ml.

Mixing: Warm white petrolatum and mineral oil in the top of a double boiler. When thoroughly mixed, turn off heat and add isopropyl alcohol. Cool and pour into jars.

Use: Apply with a tissue and wipe off.

CAUTIONS: Isopropyl alcohol may be toxic by ingestion and is flammable.

Notes: For discussion on grades of alcohol, see Appendix 4, *Denatured Alcohol.*

LIQUID MASCARA

Ingredients:
Tincture of Benzoin, 25%	1/4 C.	59 ml.
Black Dye, Oil-base	to suit	

Mixing: Mix an oil-base black dye into tincture of benzoin to reach desired color. Store in a jar.

Use: Apply with a small, soft brush.

CAUTION: As with other personal care products, if there is a possibility of an allergic reaction, apply formula first to a small test area.

Notes:

ASTRINGENT MOUTHWASH

Ingredients:
Menthol Crystals	1 speck	0.3 g.
Cinnamon Oil	to suit	
Denatured Alcohol	1/2 C.	118.0 ml.
Water	1 qt.	1.0 L.
Sodium Bicarbonate	2 T.	28.0 g.
Borax	2 T.	28.0 g.
Zinc Chloride	1/8 t.	0.5 g.
Glycerin	1 C.	237.0 ml.
Red Food Coloring	to suit	

Mixing: Mix menthol crystals and cinnamon oil into denatured alcohol, then add this mixture to water. Then add remaining ingredients, stir well, and bottle.

Use: Use as you would a standard mouthwash.

CAUTIONS: Zinc chloride is toxic. Denatured alcohol may be toxic if taken internally and is flammable.

Notes: For proper grade of alcohol, see Appendix 4, *Denatured Alcohol.*

GERMICIDAL MOUTHWASH

Ingredients:			
Oxyquinoline Sulfate	1 t.	4 g.
Water	3 T.	45 ml.
Peppermint Oil	1 t.	5 ml.
Denatured Alcohol	1/4 C.	59 ml.

Mixing: Dissolve oxyquinoline sulfate in water, then add peppermint oil and denatured alcohol. Bottle.

Use: Use full-strength for best results.

CAUTIONS: Oxyquinoline sulfate is moderately toxic in concentrated form. Denatured alcohol may be toxic by ingestion and is flammable.

Notes: For proper grade of alcohol, see Appendix 4, *Denatured Alcohol.*

MOUTHWASH POWDER

Ingredients:			
Sodium Perborate	1 C.	227.0 g.
Cornstarch	2 T.	28.0 g.
Peppermint Oil	15 Drops	1.5 ml.

Mixing: Stir all ingredients together. Store in a jar or plastic container.

Use: Add 2 teaspoons (8 g.) to 1 cup (237 ml.) of warm water. Stir and gargle.

Notes:

SWEDISH FORMULA MOUTHWASH

Ingredients:			
	Borax	1/16 t.	0.3 g.
	Boric Acid	1/2 T.	7.0 g.
	Cloves, Tincture of	to suit	
	Food Coloring	to suit	
	Water	1 qt.	1.0 L.

Mixing: Mix all the ingredients into water, stir well, and bottle.

Use: Use as you would a standard mouthwash.

CAUTION: Boric acid is poisonous if swallowed in undiluted form. Label ingredient and keep out of the reach of children.

Notes:

MOUSTACHE POMADE

Ingredients:			
	Refined Tallow	2 oz.	56 g.
	Ceresin Wax, White	3/4 oz.	21 g.
	White Beeswax	1/2 oz.	14 g.
	Mineral Oil, White	2 oz.	60 ml.
	Perfume as desired	to suit	

Mixing: Melt refined tallow, ceresin wax, beeswax, and mineral oil together in the top of a double boiler while stirring. When cool, add perfume. Stir thoroughly and bottle.

Use: Apply to manage the moustache.

Notes:

NICOTINE STAIN REMOVER

Ingredients:			
	Beeswax	1 T.	14 g.
	Paraffin Wax	1/2 T.	7 g.
	Mineral Oil	5 T.	75 ml.
	Pumice Powder	1 T.	14 g.
	Borax	1/2 T.	7 g.
	Water	3 T.	45 ml.

Mixing: Melt beeswax, paraffin wax, and mineral oil in the top of a double boiler. Turn off heat and add pumice powder and borax stirring constantly. Then add water and cool. Bottle.

Use: Rub into stained area with a cloth.

Notes:

VASELINE-TYPE PETROLEUM JELLY I

Ingredients:			
White Petrolatum	1/2 C.	113 g.
Paraffin Wax	3 T.	42 g.
Mineral Oil	2 C.	474 ml.

Mixing: Melt white petrolatum and paraffin wax in the top of a double boiler. Turn off heat, let cool for a few minutes, then add mineral oil, stirring well. Pour into jars, cool to room temperature, and cover.

Use: Use a small amount as a skin lubricant, to remove paint from hands.

Notes:

VASELINE-TYPE PETROLEUM JELLY II

Ingredients:			
White Petrolatum	1/4 C.	57 g.
Paraffin Wax	1 T.	14 g.
Ceresin Wax	1 T.	14 g.
Mineral Oil	1 C.	237 ml.

Mixing: Melt white petrolatum, paraffin wax, and ceresin wax in the top of a double boiler. Check the temperature. In a separate pan, heat mineral oil to 15° F. warmer than the first mixture. Turn off all heat and add the two mixtures together, stirring well.

Use: Use a small amount as a skin lubricant, to remove paint from hands.

Notes:

RUBBING ALCOHOL I

Ingredients:			
Denatured Alcohol or			
Isopropyl Alcohol	1-3/4 C.	414 ml.
Glycerin	1/4 C.	59 ml.
Water	1/2 C.	118 ml.

Mixing: Mix glycerin and water, then stir in denatured or isopropyl alcohol. Store in a closed bottle.

Use: Use as you would any rubbing alcohol.

CAUTIONS: Denatured and isopropyl alcohol may be toxic by ingestion and are flammable.

Notes: See Appendix 4, *Denatured Alcohol*, for proper grade of alcohol.

RUBBING ALCOHOL II

Ingredients: Denatured Alcohol or

Isopropyl Alcohol	2 C.	474.0 ml.
Glycerin	1/4 t.	1.2 ml.
Castor Oil	1/4 t.	1.2 ml.

Mixing: Stir glycerin and castor oil into denatured or isopropyl alcohol, and store in bottles.

Use: Use as a massaging aid.

CAUTIONS: Denatured and isopropyl alcohol may be toxic by ingestion and are flammable.

Notes: See Appendix 4, *Denatured Alcohol,* for proper grade of alcohol.

SCALP TREATMENT

Ingredients: Glycerin	1 T.	15 ml.
Bay Rum	1 C.	237 ml.

Mixing: Mix glycerin into bay rum. Store in a bottle.

Use: Rub well into scalp. This will help relieve dryness.

Notes:

BRUSHLESS SHAVING CREAM

Ingredients: Water	2 C.	474.0 ml.
Stearic Acid	1/2 C.	113.0 g.
Triethanolamine	1/8 t.	0.6 ml.

Mixing: Heat water to a simmer; stir in stearic acid until melted. Turn off heat and add triethanolamine. Cool to room temperature and pour into jars.

Use: Wet face with hot water, rub on beard, and shave.

CAUTIONS: Triethanolamine may be somewhat irritating to skin and mucous membranes.

Notes:

CUCUMBER SKIN LOTION

Ingredients:			
Cucumber Juice		4 C.	948.0 ml.
Glycerin		1 C.	237.0 ml.
Denatured Alcohol or			
Isopropyl Alcohol		1/4 C.	59.0 ml.
Benzoic Acid		1 speck	0.3 g.
Rose Oil		to suit	

Mixing: Obtain cucumber juice by slicing and pressing through a potato ricer, by running several through an electric juicer, or by mashing and straining through several layers of cheese cloth. Stir cucumber juice, glycerin, and denatured or isopropyl alcohol together, then add benzoic acid and rose oil. Store in bottle.

Use: This is an extra refreshing skin lotion for every part of the body from face to feet.

CAUTIONS: Denatured and isopropyl alcohol may be toxic by ingestion and are flammable.

Notes: For proper grade of alcohol, see Appendix 4, *Denatured Alcohol.*

LINSEED OIL SKIN LOTION

Ingredients:			
Linseed Oil, Boiled		1-1/2 T.	22.0 ml.
Boric Acid		2 t.	8.0 g.
Benzoic Acid		1 speck	0.3 g.
Glycerin		3 T.	45.0 ml.
Denatured Alcohol or			
Isopropyl Alcohol		10 T.	150.0 ml.
Water		1 qt.	1.0 L.
Perfume, Water-base		to suit	

Mixing: Into boiled linseed oil, mix boric acid and benzoic acid; add glycerin and denatured or isopropyl alcohol. Then stir this mixture into water and add a few drops of a water-base perfume. Store in a plastic squeeze bottle.

Use: Use as a general all-purpose skin lotion.

CAUTIONS: Denatured and isopropyl alcohol may be toxic by ingestion and are flammable.

Notes: See Appendix 4, *Denatured Alcohol,* for proper grade of alcohol. Linseed oil dries when exposed to air. Keep in an airtight container.

SKIN MOISTENER

Ingredients: Light Mineral Oil   1 C.   237 ml.
Oil-base Perfume   to suit

Mixing: Into the light mineral oil, add an oil-base perfume to suit. Store in a plastic squeeze bottle.

Use: Apply at night; wipe or wash off in the morning.

Notes:

BASIC SOAP

Ingredients: Caustic Soda (Lye)   13 oz.   364.0 g.
Cold Water   5 C.   1.2 L.
Tallow   6 lb.   2.7 kg.
Oil-base Perfume   to suit

Mixing: Using an iron, enamel, or Pyrex container, pour caustic soda into cold water slowly, stirring constantly.

Check the temperature of this mixture, and maintain it at 95° F. Then melt tallow over low heat until it reaches 130° F. Pour the caustic soda and water solution into the melted tallow in a slow, thin stream, stirring constantly and slowly until the consistency is like thick syrup. Then add an oil-base perfume, if desired. This must all be done carefully. Adding the caustic soda solution too quickly, or stirring too vigorously, can prevent the emulsion from forming.

Have ready a wooden or heavy cardboard box which has been lined with a dish towel. Pour the soap into the box and cover it to retain heat so it will cool slowly and evenly. After at least 24 hours, remove from the mold and cut into the desired size bars. Allow these to air-dry and cure for at least two weeks before using.

Use: Use as a general purpose soap.

CAUTIONS: Caustic soda heats on contact with water and can cause severe skin burns. Handle with care. Store in an airtight container. This mixture will heat up and is very caustic. If accidentally splashed on skin or clothes, flush with clear water for at least 15 minutes.

Notes:

CASTILE SOAP

Ingredients:			
	Denatured Alcohol	1 pt.	473 ml.
	Coconut Oil	4 T.	60 ml.
	Castor Oil	3 T.	45 ml.
	Palm Oil	3 T.	45 ml.
	Caustic Soda	1 T.	14 g.
	Caustic Potash	4 T.	56 g.
	Sodium Bicarbonate (Baking Soda)	1 t.	4 g.
	Sodium Chloride (Salt)	1/2 t.	2 g.

Mixing: Into the denatured alcohol, stir the remaining ingredients, mixing slowly and thoroughly. Then slowly add water, stirring constantly until the desired consistency is reached. (See instructions for molding under "Basic Soap.")

Use: Use as any fine skin soap.

CAUTIONS: Caustic soda and caustic potash heat on contact with water and can cause severe burns to skin. Handle with care. Store in an airtight container. Denatured alcohol is toxic by ingestion and is flammable.

Notes: See Appendix 4, *Denatured Alcohol,* for proper grade of alcohol.

SORE MUSCLE RUB

Ingredients:			
	Methyl Salicylate	1/2 C.	118.0 ml.
	Camphor	1/8 t.	0.5 g.
	Menthol	1/8 t.	0.5 g.
	Denatured Alcohol or		
	Isopropyl Alcohol	1-1/2 C.	355.0 ml.
	Mineral Oil	1/4 C.	59.0 ml.

Mixing: Dissolve methyl salicylate, camphor, and menthol in denatured or isopropyl alcohol, then stir in mineral oil. Store in a jar or plastic container.

Use: Apply as a general muscle rub.

CAUTIONS: Methyl salicylate is highly toxic by ingestion in concentrated form. Camphor vapors are flammable. Denatured and isopropyl alcohol is toxic by ingestion and are flammable.

Notes: See Appendix 4, *Denatured Alcohol,* for proper grade of alcohol.

TOILETTE WATER

Ingredients:		
Isopropyl Alcohol	10 T.	150.0 ml.
Glycerin	10 T.	150.0 ml.
Borax	4 T.	56.0 g.
Water	2-1/2 qt.	2.4 L.
Water-base Perfume	to suit	

Mixing: Mix isopropyl alcohol, glycerin, and borax into water. Then stir in a water-base perfume. Bottle.

Use: Splash on after bath, or for refreshment any time.

CAUTIONS: Isopropyl alcohol may be toxic by ingestion and is flammable.

Notes: For discussion on proper grades of alcohol, see Appendix 4, *Denatured Alcohol.*

VAPOR INHALANT

Ingredients:		
Pine-Needle Oil	1/4 C.	59 ml.
Tincture of Benzoin	1/4 C.	59 ml.

Mixing: Stir ingredients together. Store in an airtight container.

Use: Use 1 teaspoon (5 ml.) per quart (1 L.) of water. Put directly into vaporizer containing water.

Notes:

WRINKLE LOTION

Ingredients:		
Aluminum Potassium Sulfate (Alum)	1/4 t.	1.0 g.
Zinc Sulfate	1 speck	0.3 g.
Glycerin	2 t.	10.0 ml.
Tincture of Benzoin	2 t.	10.0 ml.
Distilled Water	1 qt.	1.0 L.
Water-base Perfume	to suit	

Mixing: Stir all ingredients into distilled water, and add a small amount of a water-base perfume, if desired. Store in a plastic squeeze bottle.

Use: Apply with a cotton swab. Leave on overnight; rinse off in the morning.

Notes:

AIR PURIFIER, WICK-TYPE

Ingredients:			
Chlorophyll		3 T.	45 ml.
Denatured Alcohol or			
Isopropyl Alcohol		2 qt.	2 L.
Water		2 qt.	2 L.
Formaldehyde, 40% solution		1 C.	237 ml.

Mixing: Mix chlorophyll, denatured or isopropyl alcohol, and water. Then stir in formaldehyde.

Use: Pour in a bottle, insert a wick, and leave about 1 inch protruding out the top. Wicks used in oil or kerosene lamps are ideal and may be obtained at most hardware stores.

CAUTIONS: Isopropyl and denatured alcohol may be toxic by ingestion and are flammable. Formaldehyde is highly toxic by ingestion, inhalation, and skin contact.

Notes: For proper grade of alcohol, see Appendix 4, *Denatured Alcohol.*

BAKING PAN ANTISTICK I

Ingredients:			
Soy Flour		1 C.	227 g.
Shortening		1 C.	227 g.
Salad Oil		2 C.	474 ml.

Mixing: Using an eggbeater, whip soy flour and shortening together, then stir in salad oil. Store in a glass bottle.

Use: Wipe pan lightly with this mixture before doing any baking.

Notes:

BAKING PAN ANTISTICK II

Ingredients: Soy Flour   2 T.   28 g.

             Wheat Flour   6 T.   84 g.

             Shortening   1 C.   227 g.

Mixing: Using a fork, cream all ingredients together. Store in a jar or can.

Use: Coat pans lightly before baking.

Notes:

BAKING-PAN SCORCH REMOVER

Ingredients: Borax   1 C.   227 g.

             Water, hot   1 gal.   4 L.

Mixing: Mix ingredients and stir. Soak pan for 30 minutes, then rinse with clear water.

Notes:

BAKING POWDER I

Ingredients: Sodium Bicarbonate (Baking Soda)   7/8 C.   198 g.

             Monocalcium Phosphate

                 (Superphosphate)   1 C.   227 g.

             Cornstarch   7/8 C.   198 g.

Mixing: Stir ingredients together with a fork.

Use: Keep in an airtight can and use wherever called for in a baking recipe.

Notes:

BAKING POWDER II

Ingredients:			
Sodium Bicarbonate (Baking Soda)	1 C.	227 g.
Sodium Aluminum Sulfate	1 C.	227 g.
Cornstarch	1 C.	227 g.

Mixing: Stir ingredients together with a fork.

Use: Keep in an airtight can and use wherever called for in a baking recipe.

Notes:

BAKING POWDER FOR DIABETICS

Ingredients:			
Powdered Casein	1/2 C.	113 g.
Monocalcium Phosphate (Superphosphate)	7 T.	98 g.
Sodium Bicarbonate (Baking Soda)	5 T.	70 g.

Mixing: Mix ingredients together with a fork.

Use: Keep in an airtight can and use wherever called for in a baking recipe.

Notes:

CHOCOLATE BISCUIT COATING

Ingredients:			
Margarine	7 T.	98.0 g.
Cocoa Powder	3 T.	42.0 g.
Dry Skim Milk	1 T.	14.0 g.
Powdered Sugar	9 T.	126.0 g.
Salt	1/4 t.	1.0 g.
Vanilla Extract	5 Drops	0.5 ml.

Mixing: Cream all ingredients together.

Use: Spread on top of biscuits for a chocolate coating.

Notes:

LEMON BISCUIT COATING

Ingredients:		
Margarine	7 T.	98 g.
Dry Skim Milk	2 T.	28 g.
Powdered Sugar	11 T.	154 g.
Yellow Food Coloring	to suit	
Lemon Oil	1 t.	5 ml.

Mixing: Cream all ingredients together.

Use: Spread as a biscuit topping.

Notes:

ORANGE BISCUIT COATING

Ingredients:		
Margarine	7 T.	98 g.
Dry Skim Milk	2 T.	28 g.
Powdered Sugar	11 T.	154 g.
Orange Food Coloring	to suit	
Orange Oil	1 t.	5 ml.

Mixing: Cream all ingredients together.

Use: Spread as a biscuit topping.

Notes:

BISCUIT FLOUR

Ingredients:		
Wheat Flour	5 lbs.	2.3 kg.
Sodium Bicarbonate (Baking Soda)	3 T.	42.0 g.
Monocalcium Phosphate (Superphosphate)	4 T.	56.0 g.
Salt	4 T.	56.0 g.
Dry Skim Milk	6 T.	84.0 g.
Shortening	2 C.	454.0 g.

Mixing: Mix all the ingredients except shortening. Then add shortening and stir in well.

Use: Use as you would a prepared biscuit flour

Notes:

BOTTLE-WASHING LIQUID I

Ingredients:			
	Caustic Soda	5 T.	70.0 g.
	Sodium Aluminate	1 t.	4.0 g.
	Water	1 gal.	3.8 L.

Mixing: Mix caustic soda and sodium aluminate into water by stirring. Rubber gloves should be worn while mixing and using this formula.

Use: Use wherever bottles, jars, and glassware must be extra-clean. This is excellent for institutional or industrial use, or for washing jars for home canning. Rinse well.

CAUTIONS: Caustic soda heats on contact with water and can cause severe burns to skin. Handle with care. Store in an airtight container.

Notes:

BOTTLE-WASHING LIQUID II

Ingredients:			
	Caustic Soda	5 T.	70.0 g.
	Sodium Carbonate (Soda Ash)	4 T.	56.0 g.
	Water	1 gal.	3.8 L.

Mixing: Mix caustic soda and sodium carbonate into water by stirring. Rubber gloves should be worn while mixing and using this formula.

Use: Use wherever bottles, jars, and glassware must be extra-clean. This is excellent for institutional or industrial use, or for washing jars for home canning.

CAUTIONS: Caustic soda heats on contact with water and can cause severe skin burns. Handle with care. Store in an airtight container.

Notes:

CELERY SALT

Ingredients:			
	Salt	7 T.	98.0 g.
	Ground Celery Seed	3 T.	42.0 g.
	Calcium Phosphate	1 speck	0.3 g.

Mixing: Dry salt in oven at slow heat for 1 hour. Then mix in ground celery seed and calcium phosphate.

Use: This is excellent for seasoning meatloaves, salads, and soups.

Notes:

CELERY VINEGAR

Ingredients:

Ethyl Alcohol (Vodka)	1 pt.	473 ml.
Tarragon Oil	1 T.	15 ml.
Celery Seed Oil	3 T.	45 ml.

Mixing: Into vodka, stir tarragon oil and celery seed oil and store in capped bottles. Let it season for at least six weeks before using.

Use: Use for a special tang in salad dressings, chili, or wherever you fancy.

Notes:

CHEESE-COATING WAX

Ingredients:

Beeswax	4 C.	908 g.
Paraffin Wax	4 C.	908 g.

Mixing: Melt ingredients in the top of a double boiler.

Use: While melted, tie a string around cheese and dip it in the wax several times, letting each coat harden before dipping again. If cheese is to be stored, this is an excellent way of protecting it.

Notes:

COATING FOR SMOKED MEATS

Ingredients:

Gelatin	3 C.	681.0 g.
Glucose	4 C.	908.0 g.
Water	1-1/2 qts.	2.5 L.

Mixing: Place all ingredients in a large pan, heat slowly to 140° F., and cook for 1-1/2 hours. Do not let temperature exceed 150° F. at any time.

Use: Dip meat in mixture or paint it on. Let dry and begin smoking immediately.

Notes:

CURRY POWDER

Ingredients:			
Coriander	1 T.	14 g.
Tumeric	1 T.	14 g.
Cinnamon	7 T.	98 g.
Cayenne	1 T.	14 g.
Mustard Seed	3 T.	42 g.
Ginger	3 T.	42 g.
Allspice	2 t.	8 g.
Fennugreek	7 T.	98 g.

Mixing: All ingredients should be ground, then stirred together well. Store in opaque container. Light is more destructive of spices than air is.

Use: Use in recipes where curry powder is called for.

Notes:

"Keep going! I'll run out and mortgage the house and meet you at the check-out."

DEFEATHERING POULTRY AND GAME BIRDS

Ingredients:	Paraffin Wax	1 lb.	454 g.
	Montan Wax	1/2 lb.	227 g.

Procedure: Melt ingredients together in the top of a double boiler. Fill a pail 1/2-full of boiling water and pour the melted wax on top of the water—it will float on the surface. Remove the larger wing and tail feathers from the bird, as well as the head. Hold the bird by the legs and immerse in the pail, pulling it out slowly so that the wax will penetrate the feathers against their natural direction. Hang the bird up to allow the wax to cool and harden, then pull the wax off and the feathers will come with it. The wax may be melted again and poured through a strainer to remove the feathers, and reused.

Notes:

DISH-AND GLASS-CLEANING POWDER

Ingredients:	Trisodium Phosphate	1 C.	227 g.
	Sodium Metaphosphate	1-1/4 C.	284 g.
	Caustic Soda	2 t.	8 g.

Mixing: Mix ingredients together with a fork. Rubber gloves should be worn while mixing and using this formula.

Use: Use 1/4 cup (57 g.) for a sinkfull of hot water.

CAUTIONS: Trisodium phosphate is a skin irritant and moderately toxic by ingestion. Caustic soda heats on contact with water and can cause severe skin burns. Handle with care. Store in an airtight container. If caustic soda comes in contact with skin, flush area with water.

Notes:

AUTOMATIC DISHWASHING DETERGENT

Ingredients:			
Soda Ash	1 C.	227 g.
Sodium Metasilicate	2 C.	454 g.
Trisodium Phosphate	1 C.	227 g.

Mixing: Stir ingredients together with a fork. Store in an airtight container.

Use: Use 1/4 cup (57 g.) in average-size automatic dishwasher.

CAUTIONS: Trisodium phosphate is a skin irritant and moderately toxic by ingestion. Use rubber gloves in handling this chemical.

Notes:

HAND DISHWASHING DETERGENT

Ingredients:		
Sodium Metaphosphate	1 C.	227 g.
Trisodium Phosphate	3 C.	681 g.

Mixing: Stir ingredients together with a fork. Store in an airtight container.

Use: Use 2 teaspoons (8 g.) for a sinkfull of hot water. It softens water and leaves dishes spotless.

CAUTIONS: Trisodium phosphate is a skin irritant and moderately toxic by ingestion. Rubber gloves should be worn in handling this chemical and while using the formula.

Notes:

DISHWASHING WATER SOFTENER

Ingredients:			
Sodium Metaphosphate	2 C.	454 g.
Sodium Metasilicate	2 C.	454 g.
Trisodium Phosphate	3/4 C.	170 g.

Mixing: Stir ingredients together with a fork. Store in an airtight container.

Use: Use approximately 1 teaspoon (4 g.) per sinkfull, depending on the hardness of the local water.

CAUTIONS: Trisodium phosphate is a skin irritant and moderately toxic by ingestion. Rubber gloves should be worn in handling this chemical and while using the formula.

Notes:

DONUT GLAZE

Ingredients:			
Water	3/4 C.	177 ml.
Corn Syrup	1 T.	15 ml.
Gelatin	1/2 t.	2 g.
Powdered Sugar	2 C.	454 g.
Any Special Flavor	to taste	

Mixing: Into water dissolve corn syrup and gelatin, by boiling. Add powdered sugar and your favorite flavoring and cool.

Use: Dip donuts in this glaze and let dry.

Notes:

EGG PRESERVATIVE I

Ingredients:			
Water	2 gal.	7.6 L.
Sodium Silicate (Waterglass)	1 qt.	1.0 L.

Mixing: Boil water and cool. Then stir in sodium silicate.

Use: Put eggs in a crock, or other large container, and cover with this solution. Keep in a cool, dark place, covered. Will keep up to six months depending upon condition of eggs when preserved.

CAUTIONS: Sodium silicate may be irritating and caustic to skin and mucous membranes.

Notes:

EGG PRESERVATIVE II

Ingredients:	Paraffin	1 C.	227 g.
	Beef Suet	1 T.	14 g.
	Boric Acid	4 T.	56 g.

Mixing: Melt paraffin and beef suet in the top of a double boiler. Turn off heat and stir in boric acid. Pour in a small cardboard container, such as a milk carton with the top cut down.

Use: Rub this compound over the shells of fresh eggs with a soft cloth, making sure to fill all the pores.

Notes:

ALMOND EXTRACT

Ingredients:	Almond Oil	1/2 t.	2.5 ml.
	Ethyl Alcohol (Vodka)	1/2 C.	118 ml.
	Water	1/2 C.	118 ml.

Mixing: Stir all the ingredients together. Store in a jar or an airtight container.

Use: Use wherever almond extract flavoring is needed.

CAUTIONS: Almond oil vapors are toxic.

Notes:

ANISE EXTRACT

Ingredients:	Oil of Anise	1 T.	15 ml.
	Ethyl Alcohol (Vodka)	1 C.	237 ml.
	Water	2 C.	474 ml.

Mixing: Stir all the ingredients together. Store in a jar or an airtight container.

Use: Use wherever anise extract flavoring is needed.

Notes:

APPLE EXTRACT

Ingredients:			
Oil of Apple	1 T.		15 ml.
Ethyl Alcohol (Vodka)	1 C.		237 ml.
Water	2 C.		474 ml.

Mixing: Stir all the ingredients together. Store in a jar or an airtight container.

Use: Use wherever apple extract flavoring is needed.

Notes:

APRICOT EXTRACT

Ingredients:			
Oil of Apricot	1 T.		15 ml.
Ethyl Alcohol (Vodka)	1 C.		237 ml.
Water	2 C.		474 ml.

Mixing: Stir all the ingredients together. Store in a jar or an airtight container.

Use: Use wherever apricot extract flavoring is needed.

Notes:

BANANA EXTRACT

Ingredients:			
Banana Oil	1 T.		15 ml.
Ethyl Alcohol (Vodka)	1 C.		237 ml.
Water	2 C.		474 ml.

Mixing: Stir all the ingredients together. Store in a jar or an airtight container.

Use: Use wherever banana extract flavoring is needed.

Notes:

CARAWAY EXTRACT

Ingredients: Oil of Caraway   1 T.   15 ml.
     Ethyl Alcohol (Vodka) 1 C.   237 ml.
     Water   2 C.   474 ml.

Mixing: Stir all the ingredients together. Store in a jar or an airtight container.

Use: Use wherever caraway extract flavoring is needed.

Notes:

CELERY EXTRACT

Ingredients: Oil of Celery   1 T.   15 ml.
     Ethyl Alcohol (Vodka) 1 C.   237 ml.
     Water   2 C.   474 ml.

Mixing: Stir all the ingredients together. Store in a jar or an airtight container.

Use: Use wherever celery extract flavoring is needed.

Notes:

CHERRY EXTRACT

Ingredients: Oil of Cherry   1 T.   15 ml.
     Ethyl Alcohol (Vodka) 1 C.   237 ml.
     Water   2 C.   474 ml.

Mixing: Stir all the ingredients together. Store in a jar or an airtight container.

Use: Use wherever cherry extract flavoring is needed.

Notes:

CINNAMON EXTRACT

Ingredients:	Cinnamon Oil	1 T.	15 ml.
	Ethyl Alcohol (Vodka)	1 C.	237 ml.
	Water	2 C.	474 ml.

Mixing: Stir all the ingredients together. Store in a jar or an airtight container.

Use: Use wherever cinnamon extract flavoring is needed.

Notes:

COFFEE EXTRACT I

Ingredients:	Finest Ground Roast Coffee	1 C.	227 g.
	Ethyl Alcohol (Vodka)	1 qt.	1 L.

Mixing: Put coffee and vodka in a pan and heat, but do not boil. Stir occasionally. When quite hot, but not boiling, pour through a paper coffee filter, and cool. Store in a jar or an airtight container.

Use: Use wherever a coffee flavor is called for.

Notes:

COFFEE EXTRACT II

Ingredients:	Water	3 pt.	1.4 L.
	Finest Ground Coffee	1 C.	227.0 g.

Mixing: Heat water to a rolling boil, stir in coffee, and turn off heat. Let stand for 30 minutes, pour through a paper coffee filter, and cool. Store in a jar or an airtight container.

Use: Use wherever a coffee flavor is called for.

Notes:

GRAPE EXTRACT

Ingredients: Oil of Grape — 1 T. — 15 ml.

Ethyl Alcohol (Vodka) — 1 C. — 237 ml.

Water — 2 C. — 474 ml.

Mixing: Stir all the ingredients together. Store in a jar or an airtight container.

Use: Use wherever a grape extract flavoring is needed.

Notes:

LEMON EXTRACT

Ingredients: Oil of Lemon — 1 t. — 5 ml.

Ethyl Alcohol (Vodka) — 1 C. — 237 ml.

Mixing: Stir the ingredients together. Store in a jar or an airtight container.

Use: Use wherever a lemon extract flavoring is needed.

Notes:

ORANGE EXTRACT

Ingredients: Oil of Orange — 1 t. — 5 ml.

Ethyl Alcohol (Vodka) — 1 C. — 237 ml.

Mixing: Stir all the ingredients together. Store in a jar or an airtight container.

Use: Use wherever orange extract flavoring is needed.

Notes:

PINEAPPLE EXTRACT

Ingredients: Oil of Pineapple 1 T. 15 ml.
Ethyl Alcohol (Vodka) 1 C. 237 ml.
Water 2 C. 474 ml.

Mixing: Stir all the ingredients together. Store in a jar or an airtight container.

Use: Use wherever a pineapple extract flavoring is needed.

Notes:

ROOT BEER EXTRACT

Ingredients: Oil of Sassafras 1 T. 15 ml.
Ethyl Alcohol (Vodka) 1 C. 237 ml.
Water 2 C. 474 ml.

Mixing: Stir all the ingredients together. Store in a jar or an airtight container.

Use: Use wherever root beer extract flavoring is needed.

Notes:

STRAWBERRY EXTRACT

Ingredients: Oil of Strawberry 1 T. 15 ml.
Ethyl Alcohol (Vodka) 1 C. 237 ml.
Water 2 C. 474 ml.

Mixing: Stir all the ingredients together. Store in a jar or an airtight container.

Use: Use wherever strawberry extract flavoring is needed.

Notes:

VANILLA EXTRACT

Ingredients: Oil of Vanilla   1 T.   15 ml.

Ingredients:		
Oil of Vanilla	1 T.	15 ml.
Ethyl Alcohol (Vodka)	1 C.	237 ml.
Water	2 C.	474 ml.

Mixing: Stir all the ingredients together. Store in a jar or an airtight container.

Use: Use wherever vanilla extract flavoring is needed.

Notes:

FIRMING TOMATOES BEFORE CANNING

Ingredients:		
Calcium Chloride	5 T.	70.0 g.
Water	1 gal.	3.8 L.

Mixing: Dissolve calcium chloride in water.

Use: After peeling, immerse tomatoes for 2 minutes.

Notes:

ICE-CUBE EASY RELEASE

Ingredients:		
Corn Oil	2 T.	30 ml.
Peanut Oil	2 T.	30 ml.

Mixing: Stir ingredients together.

Use: Apply lightly to metal surfaces of trays and dividers. Rub off excess with a cloth. One application lasts for many freezings, but do not wash trays in hot water.

Notes:

JAR AND BOTTLE SEALER

Ingredients: Gelatin   1 t.   4 g.
             Water   1-1/2 C.   355 ml.
             Glycerin   3 T.   45 ml.

Mixing: Dissolve gelatin in heated water. Add glycerin.

Use: Brush lightly around the tops of jars or bottles before putting on the lid or cap.

Notes:

JAR-SEALING WAX

Ingredients: Paraffin Wax   1 lb.   454 g.
             Stearic Acid   4 T.   56 g.

Mixing: Melt paraffin wax and stearic acid in the top of a double boiler until paraffin wax is melted.

Use: Pour the melted mixture over preserves.

Notes: The stearic acid increases the density of wax, giving a tighter seal and better protection.

JELLY COAGULANT

Ingredients: Tartaric Acid (Cream of Tartar)   3/4 C.   170 g.
             Citric Acid   1/4 C.   57 g.
             Water   2 C.   474 ml.

Mixing: Mix tartaric acid and citric acid into water, stirring well. Store in an airtight container.

Use: Use 1 to 2 tablespoons (15-30 ml.) for 7 to 8 pounds (3.2-3.6 kg.) of jelly. Adjust the proportions according to the desired consistency.

Notes:

MALTED-MILK POWDER

Ingredients:	Powdered Malt Extract	10 oz.	280 g.
	Powdered Skim Milk	4 oz.	112 g.
	Cane Sugar	6 oz.	168 g.

Mixing: Dry-mix ingredients thoroughly and transfer to an airtight container.

Use: Use as a topping over ice cream or in making malted-milk.

Notes:

MEATLOAF GLAZE

Ingredients:	Water	2 C.	474 ml.
	Sugar	1-1/4 C.	284 g.
	Paprika	5 T.	70 g.
	Glucose	10 T.	150 ml.
	Gelatin	10 T.	140 g.

Mixing: Into water, pour all ingredients, bring to a boil, and stir.

Use: When cool, brush on surface of meatloaf.

Notes:

MUSTARD

Ingredients:	Vinegar	1 C.	237.0 ml.
	Ground Mustard Seed	1/2 C.	113.0 g.
	Pepper	1 speck	0.3 g.
	Sugar	4 T.	56.0 g.
	Powdered Cloves	1 speck	0.3 g.

Mixing: Boil vinegar and pour over ground mustard seed. Stir until smooth and add remaining ingredients. Store in a jar or plastic squeeze bottle.

Use: This is great on hotdogs or wherever.

Notes:

OVEN-CLEANING POWDER

Ingredients:		
Trisodium Phosphate	1/2 C.	113 g.
Soda Ash	1/2 C.	113 g.
Sodium Perborate	1 C.	227 g.
Powdered Soap	2 T.	28 g.

Mixing: Stir ingredients together with a fork.

Use: Rub on with a damp cloth or sponge. Rinse off with a sponge and clean water.

CAUTIONS: Trisodium phosphate is a skin irritant and moderately toxic by ingestion. Wear rubber gloves while mixing and using this formula.

Notes:

OVEN-CLEANING SPRAY

Ingredients:		
Oven-cleaning Powder (Above formula)	1 C.	227 g.
Water	1 qt.	1 L.

Mixing: Stir oven-cleaning powder into water. Pour into a plastic spray bottle.

Use: Spray onto oven surface and allow to soak for 1 hour. Wipe clean with a wet cloth or sponge.

Notes:

OVEN POLISH

Ingredients:		
Soap Powder	1/4 oz.	7 g.
Water	2 oz.	60 ml.
Isopropyl Alcohol	1 oz.	30 ml.
Chromic Oxide	4 oz.	112 g.
Talc	2-1/2 oz.	70 g.

Mixing: Dissolve soap powder in water. Add isopropyl alcohol while stirring. Then stir in chromic oxide and talc. Store in an airtight container.

Use: Apply to metal surfaces of oven with a damp cloth.

CAUTIONS: Chromic oxide is toxic by ingestion. Isopropyl alcohol is flammable and may be moderately toxic by ingestion.

Notes: For discussion on grades of alcohol, see Appendix 4, *Denatured Alcohol.*

POPCORN-BALL BINDER

Ingredients:			
Sugar	1-1/2 C.	340 g.
Glucose	1 C.	237 ml.
Salt	to suit	

Mixing: Heat all three ingredients together until they boil.

Use: Pour over freshly-popped popcorn and mold into balls.

Notes:

REFRIGERATOR DEODORIZER

Ingredients:			
Activated Charcoal	1/2 C.	113 g.
Calcium Carbonate (Chalk)	1 C.	227 g.
Portland Cement	2 C.	454 g.
Vermiculite	3/4 C.	170 g.
Aluminum Powder	1 T.	14 g.

Mixing: Mix all ingredients together with a fork. Add water to form a thick, creamy consistency. Pour into molds—the cardboard boxes that kitchen matches come in are perfect.

Use: When dry, remove from molds, place in an oven at 200° F. for 24 hours, cool and place in refrigerator to pick-up odors. After a week, reactivate block by heating at 400° F. for 4 hours.

Notes:

NONCAKING SALT

Ingredients:			
Salt	1 lb.	454 g.
Potassium Chloride	2 t.	8 g.

Mixing: Into salt, stir potassium chloride. Return salt to container.

Use: This is excellent for people who live in damp climates to prevent salt from caking.

Notes:

SALT FOR ICE-CREAM MAKER

Ingredients: Rock Salt   2 C.   454.0 g.
Ammonium Nitrate   1 C.   227.0 g.
Ice   10 C.   2.3 kg.

Mixing: Mix rock salt and ammonium nitrate. Add this mixture to the ice around the ice-cream bucket.

Use: This is a good mixture to aid in freezing the ice cream.

CAUTIONS: Do not store ammonium nitrate in high temperatures.

Notes:

SHORTENING

Ingredients: Rendered Beef Fat   2 lbs.   908.0 g.
Cottonseed Oil   2 qts.   1.4 L.

Mixing: Melt rendered beef fat into cottonseed oil. Cool and store in an airtight container.

Use: Use whenever shortening is called for in cooking.

Notes:

SHORTENING IMPROVER

Ingredients: Water   2 C.   474 ml.
Lecithin   1 T.   14 g.
Propylene Glycol Monstearate   1/2 C.   118 ml.

Mixing: Into water, mix lecithin and propylene glycol monstearate. Heat until it boils and stir well. Cool and store in an airtight container.

Use: Mix 3 tablespoons (45 ml.) to 1 pound (454 g.) of regular shortening.

CAUTIONS: Propylene glycol monstearate is combustible.

Notes:

SINK DISPOSAL CLEANER

Ingredients: Ice Cubes   2 Trays
Lemons   2

Use: Cut up lemons into eighths, put them down the disposal with the ice cubes on top of them, and operate the disposal until they are eliminated. Flush with cold water.

Notes:

NONCAKING SUGAR

Ingredients: Sugar   2 lbs.   908 g.
Tricalcium Phosphate   1 t.   4 g.

Mixing: Into sugar, stir tricalcium phosphate. Return sugar to container and use as needed.

Use: This is excellent for people who live in damp climates to prevent sugar from caking.

Notes:

VANILLA SUGAR

Ingredients: Ground Vanilla Beans   4 T.   56.0 g.
Powdered Sugar   3/4 C.   170.0 g.
Almond Oil   5 Drops   0.5 ml.

Mixing: Stir all ingredients together and store in an airtight can or jar.

Use: Use in baking for an especially smooth flavor.

CAUTIONS: Vapors of almond oil are toxic.

Notes:

WHIPPED CREAM IMPROVER

Ingredients:			
	Cold Water	2 T.	30 ml.
	Gelatin	1 t.	4 g.
	Chilled Cream	1 C.	237 ml.
	Sugar	2 T.	28 g.
	Vanilla Extract	to suit	

Mixing: Put 2 tablespoons cold water (30 ml.) in a pan, sprinkle in gelatin, and heat until the gelatin is dissolved. Slowly stir this gelatin mixture into chilled cream, stir in remaining ingredients and cool for 1 hour. Then whip normally. This will give a thicker, more stable whipped cream.

Notes:

ALCOHOL RESISTANT TREATMENT FOR WOOD

Ingredients: Paraffin Oil   1 qt.   1 L.
              Vinegar   1 T.   15 ml.

Mixing: Into vinegar, mix paraffin oil. Store in a bottle or jar.

Use: Apply with a rag. The vinegar emulsifies the oil for better penetration.

Notes:

ANTIRUST TOOL COATING

Ingredients: Lanolin   1 C.   227 g.
              Petrolatum   1 C.   227 g.

Mixing: Melt ingredients in the top of a double boiler.

Use: While still warm, brush a thin coat on tools, or anything metal that you wish to protect from rust.

Notes:

BOILER COMPOUND

Ingredients:			
Soda Ash		17 lb.	7.7 kg.
Trisodium Phosphate		2 lb.	0.9 kg.
Starch		24 oz.	672.0 g.
Tannic Acid		40 oz.	1.1 kg.

Mixing: Mix all the ingredients together.

Use: Add to the water in the boiler at the rate of 1 cup (227 g.) per gallon (3.8 L.) water. This keeps scale, rust, and deposits from forming in pipes, radiators, etc.

CAUTIONS: Trisodium phosphate is a skin irritant and moderately toxic by ingestion. Use rubber gloves.

Notes:

CAULKING COMPOUND

Ingredients:			
Asphalt		1-1/2 C.	340 g.
Kerosene		1-1/4 C.	296 ml.
Asbestos Powder		2 C.	454 g.

Mixing: Dissolve asphalt into kerosene, and then mix in asbestos powder to form a puttylike consistency. Store in an airtight container.

Use: Apply to windows, walls, tub, sink installations, and so on, wherever caulking is needed.

CAUTIONS: Kerosene is toxic if taken internally and flammable. **Do not inhale dust of asbestos powder.**

Notes:

DEGREASING COMPOUND

Ingredients:			
Trisodium Phosphate		6-1/2 lb.	2.9 kg.
Sodium Carbonate (Soda Ash)		2-1/2 lb.	1.1 kg.
Sodium Metasilicate		1 lb.	454.0 g.

Mixing: Dry-mix the ingredients together and store in an airtight container.

Use: Use 1 cup (227 g.) of this mix per gallon (3.8 L.) of hot water. Apply to surface to be degreased, allowing it to soak in for half an hour, and wash with clear water.

CAUTIONS: Trisodium phosphate is a skin irritant and moderately toxic by ingestion. Use rubber gloves.

Notes:

ETCHING FLUID

Ingredients:			
Zinc Sulfate	1/2 t.	2 g.
Copper Sulfate	2 T.	28 g.
Sodium Chloride (salt)	2 T.	28 g.
Water	1/2 C.	118 ml.

Mixing: Dissolve zinc sulfate, copper sulfate, and sodium chloride into water.

Use: Used to etch glass. This is safer than an acid solution, but it is slower. Always wear rubber or plastic gloves when using the solution.

CAUTIONS: Copper sulfate is highly toxic.

Notes:

EYEGLASS POLISH

Ingredients:		
Granulated Soap	7 T.	98 g.
Glycerin	3 T.	45 ml.

Mixing: Stir granulated soap into glycerin to form a paste. Store in a plastic can or glass jar.

Use: With a soft rag, apply to eyeglasses and polish. Wash off in clean warm water.

Notes:

FENCE-POST PRESERVATIVE

Ingredients:			
Copper Sulfate		5-1/2 lb.	2.5 kg.
Water		5-1/2 gal.	21.0 L.

Mixing: Mix copper sulfate into water thoroughly.

Use: Immerse the end of the post that is to be in the ground and let stand in solution for 24 hours, then air-dry. This prevents posts from rotting.

CAUTIONS: Copper sulfate is highly toxic by ingestion and a strong skin irritant.

Notes:

FILE-CLEANING COMPOUND

Ingredients:			
Sulfuric Acid		1 C.	237 ml.
Copper Sulfate		4 T.	56 g.
Borax		4 T.	56 g.
Water		2 C.	474 ml.

Mixing: Add sulfuric acid, copper sulfate, and borax to water. Mix well.

Use: Clean file of all grease and oil with a solvent or detergent. Immerse the file in bath for one half hour, remove file, and use a wire brush to clean off deposits.

CAUTIONS: Sulfuric acid is a strong irritant to skin; use rubber gloves and use in well-ventilated area. If it inadvertently makes contact with skin, flush with clear water for prolonged periods of time. Use caution in mixing with water. Always add the acid to water, **never** the reverse. It is highly toxic and a strong irritant to tissue. Copper sulfate is highly toxic.

Notes:

FLOUR PASTE

Ingredients:			
Wheat Flour		4 T.	56 g.
Cold Water		6 T.	90 ml.
Boiling Water		1-1/2 C.	355 ml.

Mixing: Mix wheat flour and cold water thoroughly and stir into boiling water. Store in an airtight container.

Use: This is a simple adhesive for gluing paper.

Notes:

FURNITURE FINISHING OIL

Ingredients:			
Paraffin Oil	2 C.	474 ml.
Turpentine	7 T.	105 ml.
Benzene	7 T.	105 ml.
Raw Linseed Oil	5 T.	75 ml.

Mixing: Stir all the ingredients together. Store in an airtight container.

Use: Use as an oil finish for furniture. This mixture penetrates the wood very well.

CAUTIONS: Turpentine and benzene are flammable. Turpentine is also toxic if taken internally. Handle both with care.

Notes:

FURNITURE FINISHING POLISH

Ingredients:			
Turpentine	1-3/4 C.	414 ml.
Mineral Oil	1-3/4 C.	414 ml.
Fine Rottenstone	1 T.	14 g.

Mixing: Mix turpentine and mineral oil. Then stir in fine rottenstone. Store in an airtight container.

Use: Apply a small amount with a soft cloth over the final coat of varnish or lacquer. Rub well and wipe dry with a clean cloth.

CAUTIONS: Turpentine is highly toxic if taken internally and flammable. Handle with care.

Notes:

GALVANIZED COATING REPAIR

Ingredients:			
Zinc, Powdered	1/2 T.	7 g.
Lead, Powdered	1/4 t.	1 g.
Tin, Powdered	6 T.	84 g.

Mixing: Stir ingredients together with a fork.

Use: Sprinkle mixture over the damaged area, heat with a torch until the metal flows.

CAUTIONS: Powdered lead is poisonous.

Notes:

GARAGE FLOOR CLEANER

Ingredients: Powdered Soap   3 C.   681 g.
     Sodium Metasilicate   6 T.   84 g.

Mixing: Stir ingredients together. Store in an airtight container.

Use: Use 1 cup (227 g.) per gallon (3.8 L.) of hot water to scrub a concrete floor.

Notes:

GASKET CEMENT

Ingredients: White Casein Glue   6 oz.   180 ml.
     Acetic Acid   3 oz.   90 ml.
     Gelatin   1 oz.   28 g.
     Water   8 oz.   240 ml.
     Shellac   1 oz.   30 ml.

Mixing: Mix white casein glue and acetic acid in a container. In a separate container, mix gelatin and water. Now pour the first mixture into the second mixture and stir thoroughly. Add shellac with continual stirring. Store in an airtight container.

Use: Paint gasket and metal surface with this solution before applying gasket.

CAUTIONS: Shellac is based on alcohol and, as such, is flammable.

Notes:

GLASS-ETCHING COMPOUND

Ingredients: Hydrofluoric Acid   6 oz.   180 ml.
     Sulfuric Acid   1 oz.   30 ml.
     Water   3 oz.   90 ml.

Mixing: Mix hydrofluoric acid and sulfuric acid and then add that mixture to water. It is very important that the acid mixture be added to the water, not the reverse.

Use: Normally, when etching is done on glass, the surface of the glass is coated with a thin layer of paraffin wax and then the design is put on, cutting through the wax down to the glass. After this step has been completed, flow the above mixture over the cut marks in the paraffin so that it will etch the glass in the correct pattern. When etched lines have become as deep as you want them, flush immediately with a saturated solution of water and baking soda to stop the acid action.

CAUTIONS: Hydrofluoric acid is highly caustic to skin and mucous membranes and is highly toxic by ingestion and inhalation. Use with extreme care. Wear rubber gloves and use in well-ventilated area. Sulfuric acid is highly toxic and a strong irritant to skin and tissue. Wear rubber gloves and use in well-ventilated area. Always add sulfuric acid to water, **never** the reverse. In the event of contact with skin with either of these materials, flush with clear water for prolonged periods of time.

Notes:

GLASS-MARKING INK—PERMANENT

Ingredients: India Ink	2 oz.	60 ml.
Sodium Silicate	11 oz.	330 ml.

Mixing: Mix the two ingredients together.

Use: Use as you would any ink for marking glass.

CAUTIONS: Sodium silicate may be irritating and caustic to skin and mucous membranes.

Notes: Proportions of India ink may be manipulated to reach desired color.

HUMIDITY INDICATOR

Ingredients: Cobalt Chloride	1/4 C.	57 g.
Water	1/4 C.	59 ml.

Mixing: Dissolve cobalt chloride in water.

Use: Paint mixture on any paper, plastic or metal surface. Low humidity will turn this substance bright blue; high humidity will turn it pink.

Notes:

JAPAN WAX EMULSION

Ingredients:			
Japan Wax	3/4 C.	170 g.
Triethanolamine	4 t.	20 ml.
Boiling Water	2-1/4 C.	533 ml.

Mixing: Warm Japan wax and triethanolamine in a pan, then stir in boiling water. Continue stirring until an emulsion forms. Store in an airtight container.

Use: Use as a furniture and floor polish. Apply with a soft cloth, and when dry, rub well with a clean cloth.

CAUTIONS: Triethanolamine may be somewhat irritating to skin and mucous membranes.

Notes:

JEWELRY-CLEANING COMPOUND

Ingredients:			
Trichloroethylene	6 T.	90 ml.
Kerosene	2 T.	30 ml.

Mixing: Stir ingredients together.

Use: Apply to jewelry with a toothbrush to get into crevices. Wash with warm water and dry.

CAUTIONS: Vapors of trichloroethylene are toxic. Use with adequate ventilation. Kerosene is toxic by ingestion and flammable.

Notes:

JEWELRY POLISH

Ingredients:			
Ferric Oxide	5 T.	70 g.
Calcium Carbonate (Chalk)	2 T.	28 g.
Water	to make paste	

Mixing: Mix together ferric oxide and calcium carbonate, then add water, stirring slowly to make a thick paste.

Use: Apply to jewelry with a cloth or toothbrush; wash off with warm water.

Notes:

LABELING MINERAL SAMPLES

Ingredients: Plaster of Paris   1 C.   227 g.
            Water   to suit

Mixing: Into plaster of Paris, mix water to form a paste of stiff consistency.

Use: Place a small amount of this mixture on the tops of mineral samples. Tap the sample on a table or hardtop bench until surface of plaster has become smooth. When plaster has dried, brush-coat surface with a transparent adhesive, such as casein glue. When dry, the result is a smooth writing surface that can be marked with a felt pen.

Notes:

LABORATORY HAND LOTION

Ingredients: Glycerin   1 C.   237 ml.
            Phenol (Carbolic Acid)   10 Drops   1 ml.

Mixing: Into glycerin, mix phenol. Store in an airtight container.

Use: Rub into hands to clean and condition. Wipe off excess with a paper or cloth towel.

CAUTIONS: Carbolic acid is highly toxic by ingestion, inhalation, and skin absorption.

Notes:

CIGARETTE LIGHTER FLUID

Ingredients: Naphtha Deodorized   2 C.   474.0 ml.
            Citronella Oil   1/8 t.   0.6 ml.

Mixing: Stir ingredients together thoroughly. Store in an airtight container.

Use: Use as a cigarette lighter fluid.

CAUTIONS: Follow label directions on naphtha closely—it is highly flammable. Citronella oil is mildly toxic if taken internally.

Notes:

LUBRICANT, DRY POWDER

Ingredients: Zinc Stearate   1 C.   227 g.

             Talc   1 C.   227 g.

Mixing: Dry-mix ingredients thoroughly. Store in an airtight container.

Use: This may be applied with a syringe or powder atomizer to sticking wooden surfaces.

Notes:

LUBRICANT FOR FINE INSTRUMENTS

Ingredients: Mineral Oil (Lightest Weight)   1 C.   237 ml.

            Aluminum Stearate   3 T.   42 g.

Mixing: Melt the two ingredients together in the top of a double boiler, stirring until a smooth solution forms.

Use: Apply to fine instruments, (microscope gears) with an eyedropper.

Notes:

MACHINE-SHOP CUTTING LUBRICANT

Ingredients: Sal Soda   8 T.   112.0 g.

            Soft Soap   1 C.   227.0 g.

            Water   1 gal.   3.8 L.

            Paraffin Oil   1 C.   237.0 ml.

Mixing: Mix sal soda and soft soap into water with stirring. Add paraffin oil in a steady stream, with constant stirring, until an emulsion forms. Store in an airtight container.

Use: Use as a cutting lubricant for machine operations.

CAUTIONS: Sal soda is moderately toxic and an irritant to mucous membranes. Wear rubber gloves when handling.

Notes:

LUBRICANT FOR SCREW AND BOLT THREADS

| Ingredients: | Powdered Graphite | 1/4 C. | 57 g. |
| | Mineral Oil | to suit | |

Mixing: Mix ingredients to a workable consistency.

Use: Dip threads into compound or apply with brush.

Notes:

LUBRICANT FOR SLIDE RULES

| Ingredients: | Tapioca Flour | 1 C. | 227 g. |
| | Zinc Stearate | 1/4 C. | 57 g. |

Mixing: Simply mix ingredients together in a bowl, using a fork. Store in a jar or plastic container.

Use: Apply sparingly to slide rule with a soft cloth.

Notes:

LUBRICANT FOR TIRE MOUNTING

Ingredients:	Water	1 qt.	1 L.
	Triethanolamine	4 oz.	112 g.
	Stearic Acid	1 lb.	454 g.

Mixing: Bring water to a rolling boil. Add remaining ingredients. Stir until the stearic acid has melted. Store in an airtight container.

Use: Apply to tire and rim when mounting a tire.

CAUTIONS: Triethanolamine may be somewhat irritating to skin and mucous membranes.

Notes:

WOOD LUBRICANT

Ingredients: Paraffin Wax   1 C.   227 g.
            Petrolatum   1 C.   227 g.
            Silicone Oil   2 T.   30 ml.

Mixing: Melt paraffin wax and petrolatum in the top of a double boiler. Cool down to just above solidification point, add silicone oil and continue to stir. Pour into jars and allow to cool.

Use: Apply as a lubricant wherever wood contacts wood, such as drawers in cabinets, dressers, etc.

Notes:

METAL DEGREASING SOLUTION

Ingredients: Linseed Oil, Boiled   4 T.   60 ml.
            Triethanolamine   1 qt.   1 L.

Mixing: Stir boiled linseed oil into the triethanlamine. Let cool and store in an airtight container.

Use: Apply to grease-coated metal surfaces, allow to soak in for about five minutes, then wipe clean.

CAUTIONS: Linseed oil dries when exposed to air. Keep in an airtight container.

Notes:

METAL PIPE UNDERGROUND COATING

Ingredients: Vermiculite Granular Insulation   to suit
            Roofing Tar (cold application)   1 gal.   3.8 L.

Mixing: Add vermiculite to roofing tar until a brushable mastic (pasty coating) results.

Use: Paint pipe before laying underground.

Notes:

MICROSCOPE MOUNTING JELLY

Ingredients:
Gelatin	1 T.	14.0 g.
Water	6 T.	90.0 ml.
Glycerin	7 T.	105.0 ml.
Carbolic Acid	1 Speck	0.3 g.

Mixing: Mix all ingredients together; warm in the top of a double boiler over boiling water for 15 minutes, stirring constantly. When cool, drain off any excess water and store in closed jars.

Use: Place a small dab on microscope slide to hold the specimen.

CAUTIONS: Carbolic acid is highly toxic by ingestion, inhalation, and skin absorption. Use with caution.

Notes:

MICROSCOPE SLIDE CLEANER

Ingredients:
Xylol	4 T.	60 ml.
Denatured Alcohol	4 T.	60 ml.
Water	2 T.	30 ml.

Mixing: Stir all the ingredients together.

Use: Apply with a soft, lintfree cloth, wiping dry.

CAUTIONS: Xylol is flammable. Denatured alcohol may be toxic by ingestion and is flammable.

Notes: For proper grade of alcohol, see Appendix 4, *Denatured Alcohol.*

MOISTURE PROTECTION

Ingredient: Silica Gel Crystals   1/2 C.   113 g.

Mixing: Sew a small bag of light material to hold the crystals and place it in the container of whatever you wish to protect—such as tools, cameras, etc. The silica may be reactivated by removing it from the bag, spreading in on a pan, and heating in a 400° F. oven for 4 hours.

Notes:

OPAQUE PARAFFIN COATING

Ingredients:			
Paraffin Wax	2 C.	454 g.
Cottonseed Oil, Hydrogenated	2 T.	30 ml.

Mixing: Heat until ingredients are melted and can be stirred together.

Use: Use for waterproofing paper. Apply lightly with a brush to one side of paper.

Notes:

PAINT BRUSH CLEANER I

Ingredients:		
Kerosene	8 oz.	240 ml.
Denatured Alcohol or		
Isopropyl Alcohol	1 oz.	30 ml.
Diglycol Oleate	4 oz.	120 ml.
Ammonia (Household)	1 oz.	30 ml.

Mixing: Mix kerosene and denatured or isopropyl alcohol, then add diglycol oleate and household ammonia with stirring. Store in an airtight container.

Use: Apply with a brush and allow to soak. Scrape if necessary and wipe with a dry cloth. Flush surface of brush with clear water.

CAUTIONS: Kerosene is toxic by ingestion and flammable. Ammonia vapors should not be inhaled. Denatured and isopropyl alcohol may be toxic by ingestion and are flammable.

Notes: For proper grade of alcohol, see Appendix 4, *Denatured Alcohol.*

PAINT BRUSH CLEANER II

Ingredients:		
Tetrasodium Pyrophosphate	2 oz.	56.0 g.
Sodium Metasilicate	2 oz.	56.0 g.
Soda Ash	3/4 oz.	21.0 g.
Water (Warm)	1 gal.	3.8 L.
Liquid Detergent	1 T.	15.0 ml.

Mixing: Dissolve tetrasodium pyrophosphate, sodium metasilicate, and soda ash into warm water, with stirring. Add liquid detergent and continue stirring.

Use: Soak dried paint brush overnight in this solution and the next day, flush with clear water.

Notes:

PAINT CLEANER EMULSION I

Ingredients:			
Water		3 pts.	1.4 L.
Trisodium Phosphate		2 oz.	56.0 g.
Diglycol Stearate		8 oz.	224.0 g.
Naphtha		2 qts.	2.0 L.

Mixing: Heat water to boiling and add trisodium phosphate—maintain this mixture just below the boiling point. Melt diglycol and naphtha in the top of a double boiler. (Caution: Naphtha is highly flammable so hold at a very low heat.) Pour slowly into the mixture of water and trisodium phosphate mixture, stirring vigorously and constantly until an emulsion is formed. Store in an airtight container.

Use: This is good for cleaning painted surface.

CAUTIONS: Trisodium phosphate is a skin irritant and moderately toxic by ingestion. Wear rubber gloves. Naphtha is highly flammable and should be used with caution.

Notes:

PAINT CLEANER EMULSION II

Ingredients:			
Water		12 oz.	360 ml.
Diglycol Stearate		2 oz.	56 g.
Kerosene		8 oz.	240 ml.

Mixing: Heat water to 150° F. Dissolve diglycol stearate in hot water. Add kerosene and stir constantly at a relatively high speed. Temperature and mixing must be maintained until emulsion is complete. Turn off heat and continue stirring until temperature drops to 95° F. Emulsion will now be stable. Pour into suitable, airtight containers.

Use: This is also for cleaning painted surfaces.

CAUTIONS: Kerosene is toxic if taken internally and flammable.

Notes:

MILDEWPROOFING PAINT

Ingredients:			
Zinc Oxide	2-1/4 C.	511 g.
Mercuric Chloride	1 T.	14 g.
Linseed Oil	1 C.	237 ml.

Mixing: Stir zinc oxide and mercuric chloride into linseed oil.

Use: Stir this mixture into 1 gallon (3.8 L.) of an oil-base paint.

CAUTIONS: Zinc oxide is poisonous if taken internally. Mercuric chloride is highly toxic by inhalation, ingestion, and absorption through skin. Use caution in handling.

Notes: Linseed oil dries when exposed to air. Keep in an airtight container.

PAINT AND LACQUER REMOVER

Ingredients:		
Caustic Soda	16 oz.	454 g.
Corn Flour	7 oz.	196 g.

Mixing: Mix caustic soda and corn flour with **wooden** spoon or fork in a wooden, ceramic or glass bowl. Package in airtight jar.

Use: Apply to painted or lacquered surfaces with brush, allow to soak and remove with scraper, 2 tablespoons in 1 pint of water is usually adequate but may be varied to suit conditions.

CAUTIONS: Caustic soda is toxic by ingestion and is extremely irritating to skin. If skin contact is made flush freely with clear water.

Notes:

PAINT AND GREASE REMOVER

Ingredients:		
Castile Soap Chips	4 oz.	112 g.
Glycerin	1 oz.	30 ml.
Trichloroethylene	1 oz.	30 ml.
Isopropyl Alcohol	1 oz.	30 ml.
Hot Water	1 qt.	1 L.

Mixing: Mix ingredients together with constant stirring until all ingredients are dissolved. After it has cooled down to room temperature, bottle. Store in an airtight container.

Use: Apply with an old, used brush to paint to be removed from surface.

CAUTIONS: Castile soap chips are toxic by ingestion and are extremely irritating to skin. If skin contact is made, flush freely with clear water. Trichloroethylene vapors are toxic. Use with adequate ventilation. Isopropyl alcohol may be toxic by ingestion and is flammable.

Notes: For discussion on grades of alcohol. see Appendix 4, *Denatured Alcohol.*

PAINT AND LACQUER REMOVER—PASTE I

Ingredients:			
Cornstarch	1 oz.	28 g.
Household Ammonia	6 oz.	180 ml.
Calcium Carbonate (Chalk)	5 oz.	140 g.
Water	8 oz.	240 ml.

Mixing: Mix first three ingredients, then add water to make a paste.

Use: Apply to surface from which paint or lacquer is to be removed with a brush and allow to soak in thoroughly.

CAUTIONS: The vapors of household ammonia may be irritating and prolonged breathing of vapors should be avoided.

Notes:

PAINT AND LACQUER REMOVER—PASTE II

Ingredients:			
Caustic Soda	2 oz.	56 g.
Water	12 oz.	360 ml.
Cornstarch	1 oz.	28 g.
Calcium Carbonate (Chalk)	2 oz.	56 g.
Fine Sawdust	3 oz.	84 g.

Mixing: Mix caustic soda into water (see caution below), then add cornstarch, calcium carbonate and sawdust, with stirring.

Use: Apply with a brush to any surface where paint or lacquer is to be removed.

CAUTIONS: Caustic soda heats on contact with water and can cause severe burns to skin. If skin comes in contact with it, flush freely with clear water. Handle with care. Store in an airtight container.

Notes:

PAINT AND LACQUER REMOVER–PASTE III

Ingredients:
Caustic Soda	2 oz.	56 g.
Water	6 oz.	180 ml.
Soap Flakes	6 oz.	168 g.
Calcium Carbonate (Chalk)	4 oz.	112 g.

Mixing: Dissolve caustic soda into water (CAUTION mixture will heat). Dissolve soap flakes into mixture with stirring, add calcium carbonate and continue to stir.

Use: Apply with old brush, flush with water.

CAUTIONS: Caustic soda heats on contact with water and can cause severe burns to the skin. If contact is inadvertently made, flush freely with clear water for prolonged periods of time. Handle with care. Store in an airtight container.

Notes:

PAINT AND LACQUER REMOVER–POWDER

Ingredients:
Hydrated Lime Powder	7 oz.	196 g.
Potassium Carbonate	7 oz.	196 g.
Corn or Potato Flour	6 oz.	168 g.

Mixing: Mix three ingredients together with a wooden fork in a wooden, ceramic, or glass bowl. Store in an airtight container.

Use: Use about 2 tablespoons (28 g.) to 1 pint (473 ml.) of water. Apply to painted or lacquered surface and allow to soak. Then scrape off.

CAUTIONS: Hydrated lime and potassium carbonate are irritants to the skin. If contact is inadvertently made, flush freely with clear water for prolonged periods of time. Potassium carbonate is also toxic if taken internally.

Notes:

PAINT AND VARNISH REMOVER I

Ingredients:
Caustic Soda	1 C.	227 g.
Caustic Potash	3/4 C.	170 g.
Calcium Carbonate (Chalk)	2 C.	454 g.
Pumice Powder	1-1/2 C.	340 g.

Mixing: Stir all ingredients together with a fork. Store in an airtight container.

Use: Take 1/2 cup (113 g.) or so of this powder and add just enough water to form the consistency of cream. Brush on and scrape off.

CAUTIONS: Caustic soda and caustic potash heat on contact with water and can cause severe skin burns. If they come in contact with skin, flush area with water. Handle with care. Store in an airtight container.

Notes:

PAINT AND VARNISH REMOVER II

Ingredients:			
Sodium Metasilicate	1 C.	227 g.
Trisodium Phosphate	1 C.	227 g.
Boiling Water	2 qt.	2 L.

Mixing: Dissolve sodium metasilicate and trisodium phosphate in boiling water. Let cool and store in an airtight container.

Use: Brush on, let stand for 10 minutes, and wipe clean.

CAUTIONS: Trisodium phosphate is a skin irritant and moderately toxic by ingestion. Use rubber gloves.

Notes:

PAINT AND VARNISH REMOVER III

Ingredients:			
Paraffin Wax	1/4 oz.	7 g.
Benzene	2 oz.	60 ml.
Amyl Acetate	3 oz.	90 ml.
Acetone	3 oz.	90 ml.
Methanol (fully-denatured alcohol)	2-1/2 oz.	75 ml.

Mixing: Melt paraffin wax until just before it starts to solidify; then add benzene. Separately mix amly acetate, acetone, and methanol. Add this mixture to the first mixture, with constant stirring. Mix in well ventilated room.

Use: Apply to surface and allow to soak. Use scraper if necessary. Flush with clear water.

CAUTIONS: Amyl acetate is toxic. Acetone and benzene are highly flammable and volatile. Denatured alcohol is toxic by ingestion and is flammable. Use all these ingredients with caution.

Notes: For proper grade of alcohol, see Appendix 4, *Denatured Alcohol.*

PAINT AND VARNISH REMOVER IV

Ingredients:			
Paraffin Wax, Powdered	1/4 oz.	7 g.	
Benzene	4 oz.	120 ml.	
Calcium Carbonate (Chalk)	4 oz.	112 g.	
Acetone	4 oz.	120 ml.	
Fully-Denatured Alcohol or			
Isopropyl Alcohol	4 oz.	120 ml.	

Mixing: Melt powdered paraffin wax and cool down to just above solidification point and stir into benzene. Separately mix calcium carbonate, acetone and denatured or isopropyl alcohol. Pour this solution into the first solution. Store in an airtight container. Mix in well ventilated room.

CAUTIONS: Acetone is extremely volatile and flammable. Denatured and isopropyl alcohol is toxic by ingestion and are flammable. Benzene is flammable.

Notes: For proper grade of alcohol, see Appendix 4, *Denatured Alcohol.*

PENETRATING OIL

Ingredients:		
Butyl Alcohol	1 T.	15 ml.
Kerosene	2 T.	30 ml.
Mineral Oil	7 T.	105 ml.

Mixing: Mix butyl alcohol and kerosene into mineral oil. Stir well. Store in an airtight container.

Use: Apply on rusted or frozen hinges, bolts, etc. Allow 5 minutes or so to soak in, and work free.

CAUTIONS: Kerosene is toxic if taken internally and flammable.

Notes:

PUTTY

Ingredients:		
Raw Linseed Oil	1/2 oz.	15 ml.
Whiting or Calcium Carbonate	1 lb.	454 g.

Mixing: Mix ingredients to a putty consistency. Store in an airtight container.

Use: When replacing window panes, one of the problems frequently encountered is that the putty does not stick as well as it should. This can be overcome by painting the wood and the edge of the glass with white casine glue and allowing it to dry. Apply the putty over the surface and it will adhere well. For a finished job, apply a thin coat of kerosene over the putty and smooth to the desired shape.

CAUTIONS: Kerosene is toxic if taken internally and flammable.

Notes: Additional quantities of raw linseed oil may be used to provide a thinner mixture.

RUST REMOVER FOR TOOLS

Ingredients:			
Ammonium Citrate Crystals	1 T.	14 g.
Water	2 C.	474 ml.

Mixing: Stir ammonium citrate crystals into water until dissolved.

Use: Soak rusty metal overnight, wipe clean, rinse in clear water, and dry. Then apply a light coat of oil or the formula for Penetrating Oil given above.

Notes:

SCULPTURE MODELING WAX I

Ingredients:			
Beeswax	1-1/4 C.	284 g.
Lard	2 T.	28 g.
Venice Turpentine	2 T.	30 ml.
Burgundy Pitch	2 T.	30 ml.

Mixing: Melt beeswax and lard in the top of a double boiler. Stir in venice turpentine and burgundy pitch. Cool.

Use: When cool, this is a pliable but stable wax.

Notes:

SCULPTURE MODELING WAX II

Ingredients:			
	Beeswax	2 C.	454.0 g.
	Lard	2 C.	454.0 g.
	Venice Turpentine	4-1/2 C.	1.1 L.
	Clay or Talc (Powdered)	1-3/4 C.	397.0 g.

Mixing: Melt beeswax and lard in the top of a double boiler. Then add venice turpentine and powered clay or talc, stirring well. Cool.

Use: When cool, this is a pliable but stable wax.

Notes:

UNDERCOAT FOR SHELLAC

Ingredients:			
	Boric Acid	1/2 t.	2 g.
	Shellac	2 T.	30 ml.

Mixing: Mix boric acid into shellac.

Use: Apply this with a brush. It dries to a hard surface and fills pores. For the second coat, use ordinary shellac or varnish.

Notes:

"Couldn't resist, could you?"

AIRCRAFT DE-ICING COMPOUND

Ingredients:			
	Granulated Soap	2 T.	28.0 g.
	Water	3 C.	711.0 ml.
	Ethylene Glycol	1/2 gal.	1.9 L.
	Gelatin	3/4 lb.	340.0 g.
	Tragacanth Gum	1 T.	14.0 g.
	Mineral Oil	1 C.	237.0 ml.

Mixing: Dissolve granulated soap in water, then stir in remaining ingredients in order shown. Store in an airtight container.

Use: Use in any liquid de-icing container on aircraft.

Notes:

AIRCRAFT ENGINE PROTECTIVE COATING

Ingredients:			
	Lubricating Oil	16 oz.	480 ml.
	Triethanolamine	3 oz.	90 ml.

Mixing: Thoroughly mix triethanolamine into lubricating oil. Store in an airtight container.

Use: Apply with a brush to metal parts of engine to be protected.

CAUTIONS: Triethanolamine may be somewhat irritating to skin and mucous membranes.

Notes:

AIRCRAFT SURFACE CLEANER

Ingredients:		
Denatured Alcohol	1 gal.	3.8 L.
Pine Oil	1 qt.	1.0 L.
Naphtha	2-1/2 qt.	2.5 L.
Soap Flakes	4 lb.	1.8 kg.
Sulfonated Castor Oil	1 lb.	454.0 g.

Mixing: Mix denatured alcohol, pine oil, and naphtha. Then stir in soap flakes and sulfonated castor oil. Store in an airtight container.

Use: Wash down aircraft exterior with this compound, then hose off.

CAUTIONS: Denatured alcohol may be toxic by ingestion and is flammable. Naphtha is highly flammable.

Notes: For proper grade of alcohol, see Appendix 4, *Denatured Alcohol.*

"Anyone here fly a 747?"

ANTIFOG COMPOUND FOR GLASS

Ingredients: Silicone Water Emulsion   1 T.   15 ml.
            Water   3 C.   711 ml.

Mixing: Stir silicone water emulsion into water.

Use: Apply a light film to glasses, windshields, etc., with a soft cloth.

Notes:

AUTO-BATTERY TERMINAL CLEANER

Ingredients: Bicarbonate of Soda (Baking Soda)   1/2 C.   113 g.
            Water   to make a paste

Mixing: Mix just enough water into the soda to make a thin paste.

Use: Apply the paste to corrosion on battery terminals. The mixture will neutralize the acid so that the accumulation can easily be wiped or brushed off.

Notes:

AUTO-BATTERY TERMINAL COATING

Ingredients: Sodium Silicate   1 C.   227 ml.
            Water   1 C.   237 ml.

Mixing: Stir ingredients together. Store in an airtight container.

Use: Paint on battery terminals to prevent corrosion.

CAUTIONS: Sodium silicate may be irritating and caustic to skin and mucous membranes.

Notes:

AUTO BODY CLEANER

Ingredients:			
Denatured Alcohol	3/4 C.	177 ml.
Water	4-1/2 C.	1 L.
Diatomaceous Earth	1 C.	227 g.
Mineral Oil	1/3 C.	79 ml.
Methyl Salicylate	1 T.	15 ml.

Mixing: Stir together denatured alcohol and water, then stir in remaining ingredients. Store in an airtight container.

Use: For heavy road grime, oil and tar as well as encrusted insects, rub in with a cloth, then wash off.

CAUTIONS: Methyl salicylate is highly toxic by ingestion in concentrated form. Denatured alcohol may be toxic by ingestion and is flammable.

Notes: For proper grade of alcohol, see Appendix 4, *Denatured Alcohol.*

AUTO BODY LIQUID WAX

Ingredients:			
Yellow Beeswax	2 T.	28 g.
Ceresin Wax	1/2 C.	113 g.
Turpentine	2 C.	474 ml.
Pine Oil	1 T.	15 ml.

Mixing: Melt yellow beeswax and ceresin wax in the top of a double boiler. Mix the two together well and take off heat. When this just begins to solidify around the edges, slowly stir in turpentine and pine oil. Pour into a can or jar and allow to cool slowly to room temperature.

Use: Wipe on lightly, let dry, and polish with a clean, dry cloth.

CAUTIONS: Turpentine is toxic if taken internally and flammable. Handle with care.

Notes:

AUTO BODY PASTE WAX

Ingredients:			
Yellow Beeswax	2 T.	28 g.
Ceresin Wax	5 T.	70 g.
Carnauba Wax	9 T.	126 g.
Montan Wax	3 T.	42 g.
Mineral Oil	2 C.	474 ml.
Turpentine	4 T.	60 ml.
Pine Oil	1 T.	15 ml.

Mixing: Melt yellow beeswax, ceresin wax, carnauba wax, and montan wax in the top of a double boiler. Turn off heat and slowly stir in mineral oil, turpentine, and pine oil. Pour in cans or jars and cool to room temperature.

Use: Wipe on lightly, let dry, and polish with a clean, dry cloth.

CAUTIONS: Turpentine is toxic if taken internally and flammable. Handle with care.

Notes:

AUTO POLISHING CLOTH

Ingredients:			
Silicone Water Emulsion	1 T.	15 ml.
Water	1 C.	237 ml.

Mixing: Stir silicone water emulsion into water. Saturate a one-foot-square flannel cloth and hang up to dry.

Use: Rub dry cloth over the exterior of the car.

Notes:

AUTO BODY SOAP

Ingredients:			
Caustic Potash Flakes	3 T.	42 g.
Water	1 C.	237 ml.
Corn Oil	2 C.	474 ml.

Mixing: Dissolve caustic potash flakes into water, and slowly stir into corn oil. More water may be added if a thinner consistency is desired. Store in an airtight container.

Use: Use 1/4 cup (59 ml.) in a bucket of water. Wash with a cloth or sponge, rinse well with clean water.

CAUTIONS: Caustic potash heats on contact with water and can cause severe burns to skin. Handle with care. Wear rubber gloves. If it should come in contact with skin, flush freely with water for prolonged periods.

Notes:

AUTO ENGINE DEGREASING COMPOUND

Ingredients:		
Trisodium Phosphate	3-1/2 C.	794 g.
Sodium Bicarbonate (Baking Soda)	1 C.	227 g.
Sodium Metasilicate	1/2 C.	113 g.

Mixing: Stir ingredients together with a fork. Store in an airtight container.

Use: Depending on the size of the area to be worked on, mix 1 cup (227 g.) or so with enough water to form a paste, brush on with a stiff brush, wait for 15 minutes and wash off with clean water.

CAUTIONS: Trisodium phosphate is a skin irritant and moderately toxic by ingestion. Use rubber gloves.

Notes:

FUEL-LINE ICE PREVENTION

Ingredients:		
Denatured Alcohol	1 qt.	1.0 L.
Lubricating Oil, Light-weight	1 t.	5.0 ml.
Pine Oil	1 t.	5.0 ml.
Triethanolamine	1/4 t.	1.2 ml.

Mixing: Into denatured alcohol, mix lubricating oil, pine oil, and triethanolamine. Store in an airtight container.

Use: Use 1 cup (237 ml.) per 10 gallons (37.9 L.) of gasoline.

CAUTIONS: Denatured alcohol may be toxic by ingestion and is flammable. Triethanolamine may be somewhat irritating to skin and mucous membranes.

Notes: For proper grade of alcohol, see Appendix 4, *Denatured Alcohol.*

GASOLINE ADDITIVE, CARBON LOOSENER

Ingredients: Paraffin   1 C.   227 g.
             Motor Oil SAE 30   1 C.   237 ml.

Mixing: Heat paraffin in the top of a double boiler until melted. Turn off heat and stir in motor oil. When cool, add to gasoline at the rate of 1 cup (237 ml.) per 5 gallons (18.9 L.).

Notes:

GASOLINE ANTIKNOCK COMPOUND I

Ingredients: Gasoline   1 qt.   1 L.
             Denatured Alcohol   1/2 C.   118 ml.
             Hydrogen Peroxide   2 T.   30 ml.

Mixing: Stir all three ingredients together well.

Use: Use 1/2 cup (48 ml.) for each 5 gallons (18.9 L.) of gasoline.

Notes:

GASOLINE ANTIKNOCK COMPOUND II

Ingredients: Benzene   1/2 C.   118 ml.
             Hydrogen Peroxide   1 T.   15 ml.
             Denatured Alcohol   1 C.   237 ml.

Mixing: Mix benzene and hydrogen peroxide into denatured alcohol. Store in an airtight container.

Use: Use 1 teaspoon (5 ml.) per 5 gallons (19 L.) of gasoline.

CAUTIONS: Benzene is flammable. Hydrogen peroxide is highly toxic in concentrated form. Denatured alcohol may be toxic by ingestion and is flammable.

Notes: For proper grade of alcohol, see Appendix 4, *Denatured Alcohol.*

GASOLINE VAPOR-LOCK PREVENTION

Ingredients: Kerosene	3/4 C.	177 ml.
Turpentine	1/4 C.	59 ml.

Mixing: Into kerosene, mix turpentine. Store in an airtight container.

Use: Use 2 tablespoons (30 ml.) per 1 gallon (3.8 L.) of gasoline.

CAUTIONS: Kerosene and turpentine are both toxic if taken internally and flammable. Handle with care.

Notes:

ANTIFREEZE RADIATOR

Ingredients: Ethylene Glycol
Water (soft if possible)

Procedure: Find the lowest temperature you will experience, then mix in the indicated proportions.

To 0° F.–2 qt. (2 L.) of ethylene glycol per 1 gal. (3.8 L.) of water.
To -10° F.–2-1/2 qt. (2.5 L.) of ethylene glycol per 1 gal. (3.8 L.) of water.
To -20° F.–3-1/2 qt. (3.5 L.) of ethylene glycol per 1 gal. (3.8 L.) of water.
To -30° F.–1 gal (3.8 L. of ethylene glycol per 1 gal. (3.8 L.) of water.

CAUTIONS: Keep family pets away from mixture. Dispose of excess with care.

Notes:

AUTOMOBILE RADIATOR CLEANER I

Ingredients: Borax	2-1/2 lb.	1.1 kg.
Warm Water	3 qt.	3.0 L.

Mixing: Stir borax into warm water.

Use: Drain the radiator. Add this mixture and fill the rest of the way with water if necessary. Run the engine for 30 minutes. Drain radiator again, flush with clean water for 5 minutes, and refill.

Notes:

AUTOMOBILE RADIATOR CLEANER II

Ingredients:			
Kerosene		8 T.	120 ml.
Paradichlorobenzene		14 T.	210 ml.
Oleic Acid		1 T.	15 ml.
Caustic Soda (Lye)		4 T.	60 ml.

Mixing: Mix kerosene, paradichlorobenzene, and oleic acid. Add to radiator water and run engine 15 minutes. Stop engine and add caustic soda. Start engine and run for 30 minutes. Drain solution and flush radiator with clean water while engine is running.

CAUTIONS: Kerosene is toxic if taken internally and is flammable. Paradichlorobenzene is moderately toxic by ingestion and is an irritant to eyes. Caustic soda heats on contact with water and can cause severe burns to skin. Store in an airtight container. Handle with care. Wear rubber gloves. If contact is made with skin, flush freely with clean water.

Notes:

RADIATOR CORROSION INHIBITOR

Ingredient: Dimethylmorpholine
Use: Mix 6 tablespoons (90 ml.) per gallon (3.8 L.) of water or antifreeze.
CAUTIONS: Dimethylmorpholine is flammable.
Notes:

RADIATOR LEAK SEALER

Ingredients:			
Sulfite Liquor		1 C.	237 ml.
Asbestos Powder		2 T.	28 g.
Water		1 C.	237 ml.

Mixing: Stir ingredients together thoroughly. Store in an airtight container.

Use: Add mixture to radiator and run engine for at least 15 minutes.

CAUTIONS: Do not inhale dust of asbestos powder.

Notes:

RADIATOR RUST REMOVER

Ingredients:	Sodium Bisulfate	1 C.	227.0 g.
	Oxalic Acid	1 C.	227.0 g.
	Water	2 gal.	7.5 L.

Mixing: Stir sodium bisulfate and oxalic acid into water.

Use: Drain radiator, fill with this mixture, and run the engine for 1 hour. Drain and flush with clean water, and refill.

CAUTIONS: Sodium bisulfate is toxic in solution and an irritant to eyes and skin. Oxalic acid is also toxic.

Notes:

RADIATOR SCALE PREVENTION

Ingredients:	Sodium Silicate	6 T.	90.0 ml.
	Trisodium Phosphate	2 T.	28.0 g.
	Water	1 gal.	3.8 L.

Mixing: Stir sodium silicate and trisodium phosphate into water. Store in an airtight container.

Use: Put total mixture in radiator and fill with water or antifreeze.

CAUTIONS: Sodium silicate may be irritating and caustic to skin and mucous membranes. Trisodium phosphate is a skin irritant and moderately toxic by ingestion. Handle with care.

Notes:

RADIATOR SCALE REMOVER

Ingredients:	Trisodium Phosphate	1 C.	227 g.
	Water	5 gal.	19 L.

Mixing: Stir trisodium phosphate into water.

Use: Drain radiator and fill with this mixture. Run the engine at a fast idle for about 15 minutes, drain, and flush with clean water. Then refill.

CAUTIONS: Trisodium phosphate is a skin irritant and moderately toxic by ingestion. Handle with care. Use rubber gloves.

Notes:

TOP CYLINDER LUBRICANT

Ingredients:		
Lubricating Oil, Light-weight	2 C.	474 ml.
Hexachlorodiphenyl Oxide	1 t.	4 g.
Orthodichlorobenzene	1/2 t.	2 g.

Mixing: Into lubricating oil, mix hexachlorodiphenyl oxide and orthodichlorobenzene. Store in an airtight container.

Use: Use 1 teaspoon (5 ml.) per gallon (3.8 L.) of gasoline.

CAUTIONS: Orthodichlorobenzene is very irritating to skin and eyes, and moderately toxic by ingestion. Handle with care.

Notes:

AUTO TOP DRESSING

Ingredients:		
Shellac	1/2 C.	118.0 ml.
Castor Oil	1 T.	15.0 ml.
Denatured Alcohol	1 gal.	3.8 L.

Mixing: Mix shellac and castor oil into denatured alcohol. Stir well and store in airtight containers.

Use: On convertible or other fabric tops, apply lightly with a brush.

CAUTIONS: Denatured alcohol may be moderately toxic by ingestion and is flammable.

Notes: For proper grade of alcohol, see Appendix 4, *Denatured Alcohol.*

WINDSHIELD CLEANER

Ingredients:		
Calcium Carbonate (Chalk)	1/2 C.	113 g.
Sodium Bicarbonate (Baking Soda)	1/4 C.	57 g.
Diatomaceous Earth	1 C.	227 g.
Water	to make paste	

Mixing: Stir calcium carbonate, sodium bicarbonate, and diatomaceous earth together with a fork and add water, stirring slowly until a paste is formed.

Use: With a soft cloth, wipe on glass surfaces to clean and polish.

Notes:

WINDSHIELD INSECT REMOVAL

Ingredients: Denatured Alcohol   1 qt.   1 L.
            Water   1 qt.   1 L.

Mixing: Stir ingredients together. Store in an airtight container.

Use: Use in the windshield-washer fluid container, and whenever needed for cleaning off windows.

CAUTIONS: Denatured alcohol may be toxic by ingestion and is flammable.

Notes: For proper grade of alcohol, see Appendix 4, *Denatured Alcohol.*

BRAKE FLUID, HYDRAULIC I

Ingredients: Castor Oil   6 C.   1.4 L.
            Acetone   4 C.   948.0 ml.

Mixing: Into castor oil, mix acetone, with stirring until solution forms. Store in an airtight container.

Use: Use in hydraulic brake-fluid reservoir.

CAUTIONS: Acetone is extremely volatile and flammable and must be handled with great care.

Notes:

BRAKE FLUID, HYDRAULIC II

Ingredients: Ethyl Acetate   1 pt.   473 ml.
            Castor Oil   1 pt.   473 ml.

Mixing: Mix ingredients thoroughly. Store in an airtight container.

CAUTIONS: Ethyl acetate is flammable, highly combustible, and mildly toxic by inhalation and skin absorption. It is irritating to eyes and skin. If skin contact should occur, flush freely with clear water.

Notes:

CHASSIS LUBRICATION OIL

Ingredients:	Paraffin Oil	1 qt.	1 L.
	Rape Seed Oil	6 T.	90 ml.

Mixing: Into paraffin oil, mix rape seed oil. Store in a plastic spray bottle.

Use: Spray on auto spring leaves, door hinges, etc. This is good for penetrating and stopping squeaks.

Notes:

LEAK SEALER FOR TIRES

Ingredients:	Latex, Liquid Emulsion	2 C.	474 ml.
	Sodium Silicate	1 C.	237 ml.

Mixing: Stir ingredients together. Store in an airtight container.

Use: Remove valve stem from tire and pour 1/4 cup (59 ml.) into the tire casing. Be sure the inside of the valve-holder is clean before replacing valve.

CAUTIONS: Sodium silicate may be caustic and irritating to skin and mucous membranes.

Notes:

"$84 for labor! Wow, what an honor! Johnny
Carson changed my plugs."

ACIDIFIER FOR SOIL

Ingredients: Sulfur Powder, wettable   1 C.   227 g.
            Ammonium Sulfate   1 C.   227 g.
            Aluminum Sulfate   1 C.   227 g.

Mixing: Mix ingredients together in a bowl, using a fork. Before applying this mixture to soil, the existing pH of the soil should be determined. Store in an airtight container.

Use: Work into soil.

Notes: Consult Appendix 1, *pH Preferences of Some Plants,* for proper pH.

ACIDIFYING MIXTURE FOR SOIL

Ingredients: Aluminum Potassium Sulfate
            (Powdered Alum)   4 T.   56 g.
            Water   1 qt.   1 L.

Mixing: Stir aluminum potassium sulfate into water.

Use: Use about 6 tablespoons (90 ml.), mixing it into soil for each pot.

Notes: Consult Appendix 1, *pH Preferences of Some Plants,* for proper pH.

ALGAE SPRAY FOR PONDS

Ingredients: Sodium Pentachlorophenate   1 gal.   3.8 L.
              Water   5 gal.   19.0 L.

Mixing: Stir sodium pentachlorophenate into water. Store in an airtight container.

Use: Spray lightly on pond surface. Repeat in four weeks if necessary.

CAUTIONS: Sodium pentachlorophenate is toxic by inhalation and ingestion. It is an irritant to eyes and skin. Use gloves when handling.

Notes:

ANT KILLER FOR LAWNS

Ingredients: Antimony Potassium Tartarate   2 T.   28 g.
              Powdered Sugar   1 C.   227 g.

Mixing: Stir ingredients together. Store in an airtight container.

"Did I say we had to plant them 4 feet deep? I meant 4 inches."

Use: Sprinkle a small amount on each anthill on the lawn.

CAUTIONS: Antimony potassium tartarate is toxic if taken internally; read all label precautions.

Notes:

ANT REPELLENT

Ingredients:			
Sassafras Leaves	2 T.	28 g.
Water	2 C.	474 ml.

Mixing: Boil ingredients together for 5 minutes, cool, and strain out leaves.

Use: Paint liquid over ant entrances and runways.

Notes:

AZALEA FERTILIZER

Ingredients:		
Ammonium Sulfate	1-1/3 C	303 g.
Superphosphate	1-1/2 C.	340 g.
Potash	9 C.	2 kg.
Sawdust	1 bushel	8 kg.

Mixing: Thoroughly mix ammonium sulfate, superphosphate and potash into sawdust. Store in an airtight container.

Use: Mix with equal parts of soil for outside potted plants.

CAUTIONS: Potash is toxic if taken internally.

Notes:

BEDBUG POWDER

Ingredients: Aluminum Potassium Sulfate
            (Powdered Alum)   3/4 C.   170 g.
            Boric Acid   2 T.   28 g.
            Salicylic Acid   2 T.   28 g.

Mixing: Stir all ingredients together with a fork.

Use: Sprinkle in infested areas. Be sure to get into all the cracks.

Notes:

BROADLEAF WEED KILLER

Ingredient: Stoddard Solvent   1 qt.   1 L.

Use: Spray stoddard solvent on broadleaves. Be sure to coat underside of leaves as well.

CAUTIONS: Stoddard solvent is moderately flammable. Always test on small area before spraying the entire plant.

Notes:

CARPENTER ANT DESTROYER

Ingredients: Paradichlorobenzene   2 C.   454 g.
            Kerosene   2 qt.   2 L.

Mixing: Stir paradichlorobenzene into kerosene until dissolved. Store in an airtight container.

Use: Spray on infested areas.

CAUTIONS: Paradichlorobenzene is moderately toxic by ingestion and an irritant to eyes. Kerosene is toxic if taken internally and flammable.

Notes:

CATERPILLAR AND INSECT TREE BANDS

Ingredients: Rosin, Powdered	1-1/2 C.	340 g.
Linseed Oil	1 C.	237 ml.
Beeswax	1 T.	14 g.

Mixing: Heat all the ingredients in the top of a double boiler until the beeswax melts and can be stirred together with the remaining mixture.

Use: Paint a band around each tree, 3 inches wide and 18 inches above the ground.

Notes: Linseed oil dries when exposed to air. Store in an airtight container.

COCKROACH POWDER

Ingredients: Borax	4 T.	56 g.
Flour	2 T.	28 g.
Cocoa Powder	1 T.	14 g.

Mixing: Stir all ingredients together with a fork.

Use: Sprinkle on runways, or wherever cockroaches appear to be nesting.

Notes:

COMPOST MAKING

Mother Nature has provided us with a closed system that is ecologically nearly perfect. A good example of this is the tree. The root system supplies water and nutrients, and when fall comes the leaves enhance the further growth of the tree during the next growing season. Unfortunately, man, having no real appreciation for this marvelous system, has come along and largely destroyed it. Instead of leaving it as nature intended, we rake up the leaves and burn them, polluting the atmosphere. But there is a way we can follow nature's plan and reap the rewards of lush growth and increased crop yields. That way is making and using compost. And, it's free.

Compost is partially decomposed organic material that can be returned to the earth to improve the productivity of the soil. It increases water-holding capacity and improves soil structure and drainage, which aids in the removal of harmful salts. The process of making compost utilizes organic waste materials, giving them a positive value. Any organic material such as newspapers, paper boxes and bags, wood shavings and sawdust, leaves, grass clippings and kitchen wastes can be used. But there are materials that can't be composted. Examples of these are glass, metal, plastics, crockery, bones and wax-coated paper.

The process of composting is to provide the preferred bacteria with conditions ideally suited to permit them to reduce the mass to an odorless substance by decomposition. Air, and the correct amount of moisture are the two most essential elements. For this reason, the compost pile should be built on the surface of the ground, and never in a pit. Following are the procedures for successful composting.

Step 1: Contain the area where your compost pile will be with a fence that permits the circulation of air and keeps the material from blowing away or being scattered by animals or birds. The fencing can be chicken wire, chain link, or snow fencing. Our preference is the snow fencing, as it is inexpensive, flexible so it can be used as a gate, and attractive.

Step 2: Next, make a flat top pile of composting material on the ground inside the fenced area. As previously mentioned, use any organic material such as kitchen wastes, paper, etc. (In our operation, we put waste materials in newspapers or paper bags before putting them on the pile, as this eliminates the problem of flies.)

Step 3: After about 12" of composting material has been spread out in the bin, cover the entire surface with about 2" to 3" of garden soil or dehydrated manure, which will provide the bacteria that causes decomposition. Then continue adding composting materials until another 12" has been built up, etc.

Step 4: The amount of moisture you add to your compost pile is very important. The pile should be damp at all times, but never wet. And it should never have an objectionable odor. But if it does, it will be because of too much moisture. This will be your sign to back off on the amount of water.

Step 5: Composting can and should be a continuous process. So it is advisable to make your fencing cover an area sufficient in size to provide two bins, with a divider of snow fence between them. After your first pile has been allowed to build up, heat, and decompose for about six weeks, it should be forked to the second bin. Thus, the top layer of the first bin will become the bottom of the second bin. Now you can start to use the top layer of the second bin while a new supply is being built up in the first bin.

Step 6: Relatively large amounts of compost are needed to improve soil and it should be applied frequently, preferably before planting time. The method we find works best is to spread up to 6" of compost over the area, and work into the soil.

Composting is a rewarding effort. Actually it's like getting something for nothing, in that wastes are converted to a usable product.

CUT FLOWER FRESHENER I

Ingredients:			
Brown Sugar	3/4 C.	170 g.
Talc	4 T.	56 g.
Yeast	2 T.	28 g.
Pine or Lime Oil	1 t.	5 ml.
Water	1 qt.	1 L.

Mixing: In order shown, add brown sugar, talc, yeast, pine or lime oil to water, stirring well.

Use: Add 2 tablespoons (30 ml.) to a vase of flowers. (Shake before using.)

Notes:

CUT FLOWER FRESHENER II

Ingredients:		
Silver Nitrate	20 Drops	2 ml.
Granulated Sugar	1 C.	227 g.

Mixing: Stir silver nitrate into sugar so that it is completely absorbed. Store in an airtight container.

Use: Use 1 teaspoon (4 g.) in each vase of water.

CAUTIONS: Silver nitrate is highly toxic and a strong irritant. Handle with care. Read label precautions before handling.

Notes:

CUT FLOWER FRESHENER III

Ingredients:		
Manganese Chloride	2 T.	28 g.
Magnesium Chloride	2 T.	28 g.
Sodium Chloride (Salt)	2 T.	28 g.
Chloral Hydrate	4 T.	60 ml.

Mixing: Stir all ingredients together well.

Use: Use 1/4 teaspoon (1 g.) in each vase of water.

CAUTIONS: Manganese chloride is slightly toxic.

Notes:

DAIRY UTENSIL CLEANER

Ingredients: Soda Ash (Sodium Carbonate)   3/4 C.   170 g.
Sodium Bicarbonate (Baking Soda)   3/4 C.   170 g.

Mixing: Stir ingredients together.

Use: Use 1 cup (227 g.) in 1 gallon (3.8 L.) of hot water for washing utensils.

Notes:

DANDELION PLANT KILLER

Ingredients: Denatured Alcohol   2 C.   474.0 ml.
Kerosene   1 gal.   3.8 L.
Furfuraldehyde   1 lb.   454.0 g.

Mixing: Mix denatured alcohol into kerosene, then add furfuraldehyde, stirring well. Store in an airtight container.

Use: Spray over area to get rid of dandelions.

CAUTIONS: Denatured alcohol may be toxic by ingestion and is flammable. Kerosene is toxic if taken internally and flammable. Furfuraldehyde is highly toxic and can be absorbed through skin.

Notes: For proper grade of alcohol, see Appendix 4, *Denatured Alcohol.*

DANDELION ROOT KILLER

Ingredients: Kerosene   1 gal.   3.8 L.
Paraffin Oil   1 C.   237.0 ml.

Mixing: Stir ingredients together. Store in an airtight container.

Use: Remove the plant from the ground and squirt a small amount of this mixture in the root hole with a pump-type oilcan.

CAUTIONS: Kerosene is toxic if taken internally and flammable.

Notes:

FERTILIZER FOR FLOWERS

Ingredients:			
Ammonium Nitrate	1-1/4 C.	284 g.
Ammonium Chloride	3 T.	42 g.
Ammonium Phosphate	1/2 C.	113 g.
Potassium Nitrate	3/4 C.	170 g.
Calcium Sulfate	3 T.	42 g.
Iron Sulfate	2 T.	28 g.

Mixing: Stir ingredients together thoroughly with a fork. Store in an airtight container.

CAUTIONS: Do not store ammonium nitrate in high temperatures. Potassium nitrate is a dangerous fire and explosion risk when subjected to shock or heating—an oxidizing agent. Handle carefully.

Notes:

FERTILIZER FOR POTTED PLANTS

Ingredients:			
Potassium Nitrate	1 C.	227.0 g.
Superphosphate	1/3 C.	76.0 g.
Ammonium Sulfate	2 T.	28.0 g.
Urea	3 T.	42.0 g.
Calcium Carbonate (Chalk)	1 Speck	0.3 g.
Fine Sand	1/2 C.	113.0 g.

Mixing: Stir ingredients together well with a fork. Store in an airtight container.

Use: Use 1 tablespoon (14 g.) in each pot of soil when first setting out plants, and once a month thereafter as a side dressing.

CAUTIONS: Potassium nitrate is a dangerous fire and explosion risk when subjected to shock or heating—an oxidizing agent. Handle carefully.

Notes:

FLY REPELLENT FOR SCREEN DOORS

Ingredients: Phenol (Carbolic Acid)   1 t.   4 g.
           Water   1 qt.   1 L.

Mixing: Mix phenol into water while stirring. Store in an airtight container.

Use: Apply to screen door with sponge, cloth, or spray.

CAUTIONS: Phenol is highly toxic by ingestion, inhalation, and skin absorption. Use with caution.

Notes:

GRAFTING WAX FOR TREES

Ingredients: Lanolin   4 oz.   112 g.
           Rosin   3.5 oz.   98 g.
           Turpentine   1 oz.   30 ml.

Mixing: Melt ingredients together in the top of a double boiler over low heat. Store in an airtight container.

Use: Apply to joint around graft to form a seal.

CAUTIONS: Turpentine is toxic if taken internally and flammable. Handle with care.

Notes:

GRASS KILLER

Ingredients: Calcium Chloride   1 lb.   454.0 g.
           Water   1 gal.   3.8 L.

Mixing: Stir calcium chloride into water.

Use: Apply to areas of unwanted grass with a sprinkling can.

Notes·

QUACK GRASS KILLER

| Ingredients: Sodium Chlorate | 1 lb. | 454.0 g. |
| Water | 1 gal. | 3.8 L. |

Mixing: Dissolve sodium chlorate in water.

Use: Spray or sprinkle on unwanted grass.

Notes:

INSECT REPELLENT

Ingredients: Citronella Oil	2 oz.	60 ml.
Camphor	1 oz.	28 ml.
Cedar Oil	1 oz.	28 ml.
Petrolatum	8 oz.	224 g.

Mixing: Mix citronella oil, camphor, and cedar oil into petrolatum, stirring thoroughly. Store in an airtight container.

"Yeah, well as far as I'm concerned, bug killer is
bug killer and you guys owe me a new
screen door."

Use: Apply to exposed areas of skin when insects become a problem.

CAUTIONS: Citronella oil is mildly toxic if taken internally. Camphor vapors are flammable, it may cause skin irritation to some individuals; test before using on large area of skin.

Notes:

JAPANESE BEETLE SPRAY

Ingredients:			
Hydrated Lime	2 lbs.	908 g.
Water	10 gal.	38 L.
Aluminum Sulfate	3/4 C.	170 g.

Mixing: Mix hydrated lime and water, then stir in aluminum sulfate.

Use: Apply with garden sprayer. Both sides of leaves must be sprayed to insure proper control.

CAUTIONS: Hydrated lime is a skin irritant.

Notes:

MANURE FERTILIZER ACTIVATOR

Ingredients:			
Dried Yeast	1 pkg.	14.0 g.
Brown Sugar	1 lb.	454.0 g.
Warm Water	1 gal.	3.8 L.
Manure	1 bushel	8.0 kg.

Mixing: Mix dried yeast and brown sugar into warm water, then pour over manure and work in with a pitchfork.

Use: Spread a 1-inch layer over garden soil before planting and work it in with a shovel. The yeast and brown sugar speed up the composting process.

Notes:

PIGEON REPELLENT

Ingredients: Isopropyl Alcohol   1 qt.   1 L.
               Oleoresin Capsicum   1 T.   15 ml.

Mixing: Mix oleoresin capsicum into isopropyl alcohol with stirring. Paint perches where pigeons congregate such as window sills, etc. This mixture is an irritant to pigeons' feet.

CAUTIONS: Isopropyl alcohol may be toxic by ingestion and is flammable.

Notes: For discussion on grades of alcohol, see Appendix 4, *Denatured Alcohol.*

HOUSE PLANT FOOD

Ingredients: Sodium Phosphate, tribasic   6 T.   84 g.
               Potassium Nitrate   1/2 C.   113 g.

Mixing: Stir ingredients together with a fork. Store in an airtight container.

Use: Dissolve 1 tablespoon (14 g.) in 1 gallon (3.8 L.) of water, and use every two weeks.

CAUTIONS: Potassium nitrate is a dangerous fire and explosion risk when subjected to shock or heating—an oxidizing agent. Handle carefully.

Notes:

HYDROPONIC PLANT FOOD

Ingredients: Potassium Nitrate   3 T.   42 g.
               Calcium Sulfate   2 T.   28 g.
               Magnesium Sulfate   2 t.   8 g.
               Monocalcium Phosphate
                  (Superphosphate)   1 T.   14 g.
               Ammonium Sulfate   1 t.   4 g.
               Water   10 gal.   38 L.

Mixing: Dissolve the first five ingredients into water. Store in an airtight container.

Use: Use as the growing medium.

CAUTIONS: Potassium nitrate is a dangerous fire and explosion risk when exposed to shock or heating—an oxidizing agent. Handle carefully.

Notes:

POTTED PLANT FERTILIZER

Ingredients:			
	Potassium Nitrate	4-1/2 oz.	126 g.
	Ammonium Sulfate	1/2 oz.	14 g.
	Urea	1 oz.	28 g.
	Superphosphate	1-1/2 oz.	42 g.
	Calcium Carbonate (Chalk)	1/4 oz.	7 g.

Mixing: Dry-mix the ingredients, adding each in the order shown. Store in an airtight container.

Use: Work in 2 tablespoons (28 g.) of mixture to each average size pot of soil.

CAUTIONS: Potassium nitrate is a dangerous fire and explosion risk when exposed to shock or heating—an oxidizing agent. Handle carefully.

Notes:

PLANT FUNGICIDE

Ingredients:			
	Copper Chloride	5 T.	70 g.
	Lye (Caustic Soda)	3/4 C.	170 g.
	Methylcellulose	2 t.	8 g.
	Talc	3/4 C.	170 g.

Mixing: Stir all ingredients together. Store in an airtight container.

Use: Mix in 2 gallons (8 L.) of water and apply with a sprayer.

CAUTIONS: Caustic soda heats on contact with water and can cause severe burns to skin. Handle with care. Store in an airtight container.

Notes:

POTATO BLIGHT POWDER

Ingredients:			
	Copper Sulfate	1 C.	227.0 g.
	Calcium Carbonate (Slaked Lime)	4 lb.	1.8 kg.

Mixing: Stir ingredients together well. Store in an airtight container.

Use: Apply with a duster.

CAUTIONS: Both ingredients are highly toxic and skin irritants. Use care in handling.

Notes:

RABBIT REPELLENT I

Ingredients: Rosin, Gum   1 C.   227 g.
            Denatured Alcohol   1 qt.   1 L.

Mixing: Dissolve rosin into denatured alcohol. Store in an airtight container.

Use: Paint around the base of trees and shrubs that are bothered by rabbits.

CAUTIONS: Denatured alcohol may be toxic by ingestion and is flammable.

Notes: For proper grade of alcohol, see Appendix 4, *Denatured Alcohol.*

RABBIT REPELLENT II

Ingredients: Hydrated Lime   2 lb.   908 g.
            Calcium Carbonate (Chalk)   1 lb.   454 g.

Mixing: Stir ingredients together well.

Use: Dust around plants to be protected.

CAUTIONS: Hydrated lime is a skin irritant.

Notes:

TERMITE CONTROL I

Ingredients: Paradichlorobenzene   1 C.   227.0 g.
            Denatured Alcohol   5 C.   1.2 L.

Mixing: Dissolve paradichlorobenzene in denatured alcohol. Store in an airtight container.

Use: Apply several coats with a brush.

CAUTIONS: Paradichlorobenzene is moderately toxic by ingestion and an irritant to eyes. Denatured alcohol may be toxic by ingestion and is flammable.

Notes: This formula is not recommended for indoor use due to the objectionable residual odor of paradichlorobenzene. For proper grade of alcohol, see Appendix 4, *Denatured Alcohol.*

TERMITE CONTROL II

Ingredients: Pentachlorophenol   1 C.   237 ml.
            Fuel Oil   3 qt.   3 L.

Mixing: Stir fuel oil into pentachlorophenol. Store in an airtight container.

Use: Apply to wood with brush or sprayer.

CAUTIONS: Pentachlorophenol is highly toxic by inhalation, ingestion, and skin absorption. Handle carefully and wash all exposed skin well after use. This solution will kill many broadleaf plants. Spray with care around vegetation.

Notes.

WHITEWASH

Ingredients: Sodium Chloride   2 lb.   908.0 g.
            Water   1 gal.   3.8 L.
            Hydrated Lime   7 lb.   3.2 kg.

Mixing: Dissolve sodium chloride in water, then slowly stir in hydrated lime. Store in an airtight container.

Use: Brush on with a wide brush to preserve outside wood fences. Keep it well mixed as you use it.

CAUTIONS: Hydrated lime is a skin irritant.

Notes:

ANTISEPTIC WHITEWASH

Ingredients: Water   3 gal.   11.5 L.
            Slaked Lime   7 lb.   3.2 kg.
            Casein Adhesive   6 oz.   168.0 g.
            Formaldehyde   6 oz.   168.0 g.

Mixing: Mix sufficient amount of water into slaked lime, while stirring to make a smooth paste. Add balance of water, stirring continually. Now add casein adhesive and formaldehyde into mixture and stir until homogenous. Store in an airtight container.

Use: This whitewash is useful around animals. For example, it is good for painting the inside of a chicken coop.

CAUTIONS: Formaldehyde is highly toxic by ingestion, inhalation, or skin contact. Use with caution. If contact is made with skin, flush freely with clear cold water.

Notes:

WILD ANIMAL REPELLENT

Ingredients:			
Paradichlorobenzene (Moth Balls)	1 lb.	454 g.
Fine Mason Sand	1 lb.	454 g.

Mixing: Dry-mix ingredients thoroughly. Store in an airtight container.

Use: Sprinkle around crops, such as sweet corn, for control of mice, rabbits, etc.

CAUTIONS: Paradichlorobenzene is moderately toxic by ingestion and an irritant to eyes.

Notes:

WOOD FENCE-POST PRESERVATIVE

Ingredients:		
Zinc Chloride	1-3/4 lb.	794.0 g.
Water	1 gal.	3.8 L.

Mixing: Dissolve zinc chloride into water. Store in an airtight container.

Use: Soak post ends in this solution for 48 hours. Treatment should extend two or three inches above what will be ground level on the posts.

CAUTIONS: Zinc chloride is toxic.

Notes:

WOUND DRESSING FOR TREES AND SHRUBS

Ingredients:	Zinc Oxide	1 C.	227 g.
	Mineral Oil	2 C.	474 ml.

Mixing: Stir zinc oxide into mineral oil. Store in an airtight container.

Use: Apply with a brush to any pruning wound or damaged area of bark to prevent infection.

CAUTIONS: Zinc oxide is toxic if taken internally.

Notes:

©1977 Universal Press Syndicate 12/30

"I spilt the stupid plant food."

DANDRUFF TREATMENT

Ingredients:			
Salicylic Acid	1 T.	14 g.
Glycerin	3 T.	45 ml.
Denatured Alcohol	1 qt.	1 L.

Mixing: Mix salicylic acid and glycerin into denatured alcohol. Store in an airtight container.

Use: Rub well into coat against the direction of the hair. Let stand for 5 minutes, then wash off.

CAUTIONS: Denatured alcohol is toxic by ingestion and is flammable.

Notes: For proper grade of alcohol, see Appendix 4, *Denatured Alcohol.*

ANIMAL DEODORANT SPRAY

Ingredients:			
Cedar Oil Emulsion	1 t.	5 ml.
Water	1 qt.	1 L.

Mixing: Stir cedar oil emulsion into water.

Use: Spray on and around animal living quarters.

Notes:

ANIMAL DRY-CLEANING POWDER

Ingredients:			
Trisodium Phosphate		2 T.	28 g.
Talc		2 T.	28 g.
Sodium Carbonate (Soda Ash)		5 T.	70 g.
Carbolic Acid		1 T.	15 ml.
Talc		14 T.	196 g.
Starch		1-1/2 C.	340 g.

Mixing: Stir all ingredients together until carbolic acid is completely absorbed and mixed in. Store in an airtight container.

Use: Rub in against the direction of the hair, and brush vigorously or vacuum to remove.

CAUTIONS: Trisodium phosphate is a skin irritant and moderately toxic by ingestion. Use rubber gloves. Carbolic acid is toxic by ingestion, inhalation, and skin absorption.

Notes:

EAR MANGE TREATMENT

Ingredients:		
Glycerin	1 C.	237 ml.
Cresol	2 T.	30 ml.
Carbolic Acid	2 T.	30 ml.

Mixing: Stir all ingredients together. Store in an airtight container.

Use: Swab inside of ears with a piece of cotton.

CAUTIONS: Cresol is toxic and an irritant. Carbolic acid is toxic by ingestion, inhalation, and skin absorption.

Notes:

ECZEMA TREATMENT

Ingredients:		
Tincture of Iodine	3 T.	45 ml.
Glycerin	3/4 C.	177 ml.

Mixing: Stir tincture of iodine into glycerin. Store in an airtight container.

Use: Apply daily with a cotton swab.

CAUTIONS: Tincture of iodine concentration may increase as alcohol evaporates. Keep tightly capped, avoiding older containers that have been previously opened; iodine can cause severe burns in its concentrated form.

Notes:

EYE WASH I

Ingredients:			
Sodium Bicarbonate (Baking Soda)	1 T.	14.0 g.
Borax	1 T.	14.0 g.
Sodium Chloride (Salt)	1 T.	14.0 g.
Glycerin	1 Drop	0.1 ml.
Water, Distilled	1 C.	237.0 ml.

Mixing: Dissolve sodium bicarbonate, borax, and sodium chloride into a mixture of the glycerin and distilled water.

Use: Put in animal's eyes with an eyedropper, wipe excess away from edges with a clean, soft cloth or tissue.

Notes:

EYE WASH II

Ingredients:			
Boric Acid	1 t.	4 g.
Water, Distilled	1 qt.	1 L.

Mixing: Stir boric acid into distilled water until dissolved.

Use: Put in animal's eyes with an eyedropper, wipe excess away from edges with a clean, soft cloth or tissue.

Notes:

FLEA SOAP

Ingredients:		
Oleic Acid	1/4 C.	59 ml.
Isopropyl Alcohol	2-1/2 C.	592 ml.
Triethanolamine	2 T.	30 ml.

Mixing: Stir oleic acid into alcohol, then add triethanolamine. Store in an airtight container.

Use: Wet animal thoroughly, then apply small amount, rubbing in well. Rinse after 5 minutes.

CAUTIONS: Isopropyl alcohol is toxic by ingestion and is flammable. Triethanolamine may be somewhat irritating to skin and mucous membranes.

Notes: Kerosene may be substituted for alcohol if dog is not allergic. **Do not use on cats.** For discussion on grades of alcohol, see Appendix 4, *Denatured Alcohol.*

FLEA SPRAY

Mix the above soap with equal parts of water and spray on infested areas.

Notes:

MANGE TREATMENT

Ingredients:		
Lauryl Pyridinium Chloride	1 T.	14 g.
Cottonseed Oil	1 pt.	473 ml.

Mixing: Mix lauryl pyridinium chloride into a small amount of cottonseed oil until dissolved, then add the rest of the cottonseed oil.

Use: Rub in with a cotton swab against the direction of the hair.

CAUTIONS: Lauryl pyridinium chloride may be mildly irritating to skin.

Notes:

SKIN ABRASION LOTION

Ingredients:			
	Soap	1/2 C.	113 g.
	Water	1 qt.	1 L.
	Tincture of Iodine	1 T.	15 ml.

Mixing: Warm soap and water together until soap is melted. Then stir in tincture of iodine and cool. Store in an airtight container.

Use: Apply to area where skin has been scraped with a cotton swab.

CAUTIONS: Tincture of iodine concentration may increase as alcohol evaporates. Keep tightly capped, avoiding older containers that have been previously opened; iodine can cause severe burns in its concentrated form.

Notes:

AQUARIUM DISINFECTANT

Ingredients:			
	Tapwater (Room Temperature)	6 gal.	23 L.
	Household bleach	1/8 C.	28 g.

Use: Put tapwater in aquarium and add household bleach. Stir thoroughly with a brush, brushing all corners and edges. Empty water and dry thoroughly before refilling with fresh water.

CAUTIONS: The fumes of chlorine should not be inhaled and it should never be taken internally.

Notes:

AQUARIUM WATER CHLORINE REMOVAL

Ingredients:			
	Sodium Thiosulfate	1 speck	1 g.
	Tapwater	6 gal.	23 L.

Mixing: Dissolve sodium thiosulfate into tapwater. Let stand in aquarium for 1 hour before putting fish in water.

Use: If the odor of chlorine can be detected in water, it's not safe for some fish. This formula will dechlorinate it to a safe level.

CAUTIONS: The concentrated chlorine vapors should not be inhaled and chlorine should never be taken internally.

Notes:

CAT LITTERBOX DEODORANT

Ingredients: Wood shavings or Commercial

Cat Litter	3 lb.	1.4 kg.
Bicarbonate of Soda	1/2 lb.	227.0 g.

Mixing: Stir ingredients together.

Use: Use in catbox. The soda will increase the absorptive ability of the litter.

Notes:

SALT BLOCKS

Ingredients: Ball Clay	25 lb.	11.5 kg.
Water	as needed	
Coarse Salt	25 lb.	11.5 kg.

Mixing: Dampen ball clay with water, mixing into the consistency of putty. Then mix coarse salt, adding more water as necessary to keep the same consistency. Pack into wooden molds and air-dry for three days.

Use: This is a salt-licking block for horses and cattle and can be used wherever necessary.

Notes:

CATTLE RINGWORM CONTROL

Ingredients: Tincture of Iodine	1 C.	237 ml.
Glycerin	1 C.	237 ml.

Mixing: Stir ingredients together. Store in an airtight container.

Use: Apply directly to skin with a cloth or piece of cotton.

CAUTIONS: Tincture of iodine concentration may increase as alcohol evaporates. Keep tightly capped; avoid using containers that have been previously opened; iodine can cause severe burns in its concentrated form.

Notes:

DEHORNING SALVE

Ingredients:	Pine Tar	10 lb.	4.5 kg.
	Tannic Acid	1 lb.	454.0 g.

Mixing: Mix ingredients together well.

Use: Apply liberally to stump of horn of the animal.

Notes:

ANIMAL BATH POWDER

Ingredients:	Powdered Soap	1-1/4 C.	284 g.
	Trisodium Phosphate	1/2 C.	113 g.
	Sodium Bicarbonate (Baking Soda)	2 T.	28 g.
	Boric Acid	1 T.	14 g.

Mixing: Stir all ingredients together well with a fork. Store in an airtight container.

Use: Use about 3 tablespoons (42 g.) per gallon (3.8 L.) of warm water to wash down dogs or other animals. Rinse cats well to prevent ingestion of boric acid by licking.

CAUTIONS: Trisodium phosphate is moderately toxic by ingestion and a skin irritant. Use rubber gloves.

Notes:

DOG COAT DRESSING

Ingredients:	Mineral Oil	1-1/2 C.	355 ml.
	Pine Oil	1/2 C.	118 ml.

Mixing: Stir ingredients together well.

Use: Dampen a cloth with this solution and rub into the coat, then brush out. This will give additional luster to a show coat.

Notes:

ANIMAL SHAMPOO

Ingredients:			
Soft Soap	1/2 C.	113 g.
Water	1-1/2 C.	355 ml.
Glycerin	2-1/2 T.	37 ml.
Carbolic Acid	1 t.	5 ml.
Denatured Alcohol	2 T.	30 ml.

Mixing: Slowly warm soft soap and water together. When soap is melted, stir in glycerin and carbolic acid. Cool and stir in denatured alcohol. Store in an airtight container.

Use: Thoroughly wet down animal, work in small amounts of this shampoo, rub in well, and rinse off.

CAUTIONS: Carbolic acid is toxic by ingestion, inhalation and skin absorption. Denatured alcohol may be toxic by ingestion and is flammable.

Notes: For proper grade of alcohol, see Appendix 4, *Denatured Alcohol.*

ANIMAL EARACHE OIL

Ingredients:			
Glycerin	3 T.	45 ml.
Almond Oil	3 T.	45 ml.

Mixing: Stir ingredients together. Store in a glass bottle.

Use: Put a few drops in the sore ear with a sterile dropper.

CAUTIONS: Almond oil vapors are toxic.

Notes:

DOG REPELLENT FOR SHRUBS AND TREES

Ingredients:			
Anhydrous Lanolin	1/2 oz.	14 g.
Amyl Mercaptan	1 oz.	30 ml.
Creosote	1/2 oz.	15 ml.
Isopropyl Alcohol	9 oz.	270 ml.

Mixing: Mix anhydrous lanolin, amyl mercaptan, and creosote into isopropyl alcohol with stirring. Store in an airtight container.

Use: Apply around base of shrubs and trees to repell dogs.

CAUTIONS: Isopropyl alcohol is toxic by ingestion and is flammable.

Notes: For discussion on grades of alcohol, see Appendix 4, *Denatured Alcohol.*

EARWAX REMOVER

Ingredients: Isopropyl Alcohol   1/4 C.   59 ml.
Glycerin   10 Drops   1 ml.

Mixing: Stir ingredients together. Store in an airtight container.

Use: Apply to inside of ear with a sterile syringe. Remove wax carefully with a cotton swab.

CAUTIONS: Isopropyl alcohol may be toxic by ingestion and is flammable.

Notes: For discussion on grades of alcohol, see Appendix 4, *Denatured Alcohol.*

HORSE HOOF GREASE

Ingredients: Mineral Oil   1 C.   237 ml.
Petrolatum   1 C.   227 g.
Paraffin Wax   1 C.   227 g.

Mixing: Heat all ingredients together in the top of a double boiler until petrolatum and paraffin wax are melted and can be mixed together; then cool and pour into containers.

Use: Apply with a stiff brush.

Notes:

LINIMENT FOR HORSES

Ingredients: Camphor   1/2 C.   113.0 g.
Tincture of Arnica   5 C.   1.2 L.
Isopropyl Alcohol   3-1/2 C.   829.0 ml.
Diglycol Oleate   1/2 C.   118.0 ml.

Mixing: Mix camphor, tincture of arnica, and isopropyl alcohol into diglycol oleate. Mix thoroughly and bottle in an airtight container.

Use: Apply to sore muscles and joints.

CAUTIONS: Camphor vapors are flammable. Tincture of arnica is toxic by ingestion. Isopropyl alcohol may be toxic by ingestion and is flammable. Diglycol oleate is combustible.

Notes:

INDELIBLE INK FOR ANIMAL TATTOOS

Ingredients:		
Isopropyl Alcohol	1 t.	5 ml.
Glycerin	4 T.	60 ml.
Carbon Black	to suit	

Mixing: Mix isopropyl alcohol and glycerin. Add carbon black until desired darkness is achieved. Store in an airtight container.

Use: Apply to animal ear for identification with suitable needle point.

CAUTIONS: Isopropyl alcohol is flammable and mildly toxic by ingestion.

Notes:

INSECT REPELLENT FOR ANIMALS

Ingredients:		
Trichloroethylene	1 C.	237 ml.
Alcohol, Isopropyl or Kerosene	1 C.	237 ml.
Castor Oil	1 T.	15 ml.
Mineral Oil	1 T.	15 ml.
Eucalyptus Oil	2 t.	10 ml.

Mixing: Mix ingredients together in order shown. Store in an airtight container.

Use: Use in a spray bottle on and around animals where they are bothered by flies and other insects. Do not use kerosene on animals which are sensitive to the solvent. Test on a small area of skin for reaction.

CAUTIONS: Trichloroethylene vapors are toxic. Use with adequate ventilation. Isopropyl alcohol may be toxic by ingestion and is flammable. Kerosene is toxic if taken internally and flammable.

Notes: For discussion on grades of alcohol, see Appendix 4, *Denatured Alcohol.*

POULTRY ANTISEPTIC INHALANT

Ingredients:		
Liquid Soap	2 C.	474 ml.
Eucalyptus Oil	5 T.	75 ml.
Pine Oil	5 T.	75 ml.
Guaiacol	10 T.	150 ml.
Camphor Oil	5 T.	75 ml.

Mixing: Stir all ingredients together. Store in an airtight container.

Use: Use 1 tablespoon (15 ml.) in 1 pint (473 ml.) of water. Spray inside of poultry house.

CAUTIONS: Guaiacol is moderately toxic.

Notes:

POULTRY LAYING MASH

Ingredients:			
	Ground Corn	9 lb.	4.0 kg.
	Ground Barley	9 lb.	4.0 kg.
	Ground Wheat	9 lb.	4.0 kg.
	Ground Oats	9 lb.	4.0 kg.
	Meat Scraps	5 lb.	2.3 kg.
	Dried Milk Solids	5 lb.	2.3 kg.
	Alfalfa Meal	2-1/2 lb.	1.2 kg.
	Bonemeal	1 lb	454.0 g
	Salt	1 lb.	454.0 g.

Mixing: Stir all ingredients together and bag.

Use: Feed chickens for better laying.

Notes:

POULTRY LICE CONTROL

Ingredients:			
	Napthalene Powder	2 C.	454 g.
	Sulfur, ground	2 C.	454 g.
	Tobacco Powder	4 C.	908 g.
	Talc	12 C.	3 kg.

Mixing: Stir all ingredients together—sift to insure thorough mixing. Store in an airtight container.

Use: Dust on fowl against direction of feathers with a dust sprayer.

Notes:

ALCOHOL SOLID FUEL, STERNO-TYPE I

Ingredients:			
Stearic Acid		5-1/2 T.	77.0 g.
Denatured Alcohol		1 gal.	3.8 L.
Caustic Soda		2 t.	9.0 g.
Water		2 T.	30.0 ml.

Mixing: Heat just stearic acid until it melts. Meanwhile heat denatured alcohol to 160° F. and gradually add the melted stearic acid to this, stirring constantly. Dissolve caustic soda into water, and stir into the combination of alcohol and stearic acid. Remove from heat and pour into small cans. Allow to cool to room temperature and cover to prevent evaporation.

Use: Uncover and ignite with a match for a heat source.

CAUTIONS: Denatured alcohol may be toxic by ingestion and is flammable. Caustic soda heats on contact with water and can cause severe burns to skin. Handle with care. Store in an airtight container. If contact is made with skin, flush freely with clean water for prolonged periods of time.

Notes: For proper grade of alcohol, see Appendix 4, *Denatured Alcohol.*

ALCOHOL SOLID FUEL, STERNO-TYPE II

Ingredients:			
Stearic Acid		1/4 C.	57 g.
Denatured Alcohol		1 qt.	1 L.
Caustic Soda		1 T.	14 g.

Mixing: Dissolve stearic acid into 1/2 (474 ml.) of denatured alcohol, and dissolve caustic soda into the balance of alcohol. Warm both solutions separately to 140° F. and mix together. Pour into cans and cool. Store in an airtight container to prevent evaporation.

Use: Uncover and ignite with a match for a heat source.

CAUTIONS: Denatured alcohol may be toxic by ingestion and is flammable. Caustic soda heats on contact with water and can cause severe burns to skin. Handle with care. Store in an airtight container. If contact is made with skin, flush freely with clean water for prolonged periods of time.

Notes: For proper grade of alcohol, see Appendix 4, *Denatured Alcohol.*

BACON PRESERVATIVE

Ingredients:		
Acetic Acid	3 T.	45 ml.
Water	1 C.	237 ml.

Mixing: Mix acetic acid into water.

Use: For camping trips, soak a cloth in this solution and wrap around the bacon to prevent mold from forming.

Notes:

"So much for camping!"

BOAT CAULKING WAX

Ingredients:			
Beeswax	4 oz.	112 g.
Beef Tallow	5 oz.	140 g.
Lard	8 oz.	224 g.

Mixing: Place all ingredients in the top of a double boiler and heat. Stir together to form a puttylike consistency.

Use: Let cool and caulk between boards on boat.

Notes:

CANVAS FIREPROOFING

Ingredients:			
Ammonium Phosphate	1/2 C.	113 g.
Ammonium Chloride	1 C.	227 g.
Water	1 qt.	1 L.

Mixing: Mix ammonium phosphate and ammonium chloride into water, stirring until dissolved.

Use: Apply to tents, back-packs, etc. Paint or spray on canvas or soak canvas in the solution. Repeat treatment after exposure to rainy weather.

Notes:

CANVAS TENT CLEANER

Ingredients:			
Calcium Hypochlorite	3 T.	42 g.
Water	1 qt.	1 L.

Mixing: Stir calcium hypochlorite into water until dissolved.

Use: Sponge on tent, rub in, and rinse off with clear water.

CAUTIONS: Calcium hypochlorite is toxic. Read and follow label precautions.

Notes:

CANVAS WATERPROOFING

Ingredients: Turpentine   1-1/2 C.   355 ml.
Soybean Oil   3 C.   711 ml.

Mixing: Stir turpentine into soybean oil. Store in an airtight container.

Use: Paint or spray on canvas. Repeat treatment after prolonged exposure to weather.

CAUTIONS: Turpentine is toxic if taken internally and flammable. Handle with care.

Notes:

DOG FOOT CONDITIONER

Ingredients: Black Tea (Cheapest Grade)   1/4 C.   57 g.
Water   1 C.   237 ml.

Mixing: Mix tea into water and boil.

Use: Before taking dogs on long hunting, fishing, or camping trips, soak the dog's feet in this solution daily for 30 days, which will toughen his pads, making it better able to withstand rough terrains and fields.

Notes:

FISH-SCALE REMOVER

Ingredients: Water   1 qt.   1 L.
Sodium Chloride (Salt)   1/2 C.   113 g.
Alum (Aluminum Potassium Sulfate)   1/2 oz.   14 g.

Mixing: Heat water to just below boiling point and dissolve sodium chloride and alum into water.

Use: Dip fish in solution until scales stand up, then remove with knife or scaling utensil.

Notes:

FISHLINE DRESSING

Ingredients:			
Anhydrous Lanolin	2 T.	28 g.
Petrolatum	2 T.	28 g.
Castor Oil	1 t.	5 ml.

Mixing: Heat all ingredients in the top of a double boiler until they can be stirred together. Cool until the mixture just starts to solidify, then pour into molds such as small cardboard matchboxes.

Use: Hold the line taut and run this over the line on both sides. Do a section at a time until the whole line is dressed. Use on superior linen or flax lines. Not for use on nylon or other monofilament lines.

Notes:

FUEL TABLETS

Ingredients:			
Paraffin Wax	3 lb.	1.4 kg.
Fine Sawdust	4 lb.	1.8 kg.
Cardboard Molds		

Mixing: For the molds, cut cardboard tubes into 2-inch sections and seal one end with a cardboard disc. Melt 2 lb. (908 g.) of the paraffin wax over a low flame. Be very careful not to get it too hot as it is flammable. Turn off heat and mix in as much sawdust as necessary to get a very thick, but still pourable, creamy consistency. Cool slightly and pour into molds. Melt the remainder of the wax, 1 lb. (454 g.), and when the molded plugs are cool, dip them in the wax to coat the outsides.

Use: Ignite the wax-impregnated cardboard mold which acts as a wick, burning the wax and sawdust inside.

Notes:

GOLF BALL CLEANER

Ingredients:			
Household Ammonia	1/4 C.	59 ml.
Water	3/4 C.	177 ml.

Mixing: Stir ingredients together.

Use: Soak balls in this solution overnight, rinse with clean water.

CAUTIONS: Do not inhale the fumes of ammonia.

Notes:

GOLF CLUB GRIP WAX

Ingredients: Yellow Beeswax   4 T.   56 g.
            Rosin   1/2 T.   7 g.

Mixing: Melt ingredients in the top of a double boiler. Cool until it just begins to solidify and pour in molds, such as the cardboard tubes from rolls of toilet tissue.

Use: Rub a small amount on hands. Peel away the cardboard as it is used up.

Notes:

GOLF CLUB HEAD CLEANER

Ingredients: Powdered Alum (Aluminum
            Potassium Sulfate)   1/4 C.   57 g.
            Trisodium Phosphate   3/4 C.   170 g.

Mixing: Stir ingredients together with a fork. Store in an airtight container.

Use: Pick up a small amount with a damp cloth or sponge and rub on clubs. Rinse off with a damp cloth.

CAUTIONS: Trisodium phosphate is a skin irritant. Use rubber gloves. It is also moderately toxic by ingestion.

Notes:

GUN-BLUING COMPOUND

Ingredients: Ferric Chloride   3/4 C.   170 g.
            Antimony Chloride   1/2 C.   113 g.
            Tannic Acid   1/2 C.   113 g.
            Water   1-1/2 C.   355 ml.
            Linseed Oil   1 T.   15 ml.

Mixing: Dissolve ferric chloride, antimony chloride, and tannic acid into water.

Use: Clean metal thoroughly. Apply two coats of above mixture with a brush or soft cloth. Allow to dry between coats, then rub down with a cloth soaked in linseed oil.

CAUTIONS: Antimony chloride is highly toxic. Read all label precautions.

Notes: Linseed oil dries when exposed to air. Store in an airtight container.

GUN-CLEANING OIL

Ingredients: Triethanolamine   1 T.   15 ml.
Denatured Alcohol   1 t.   5 ml.
Mineral Oil   1 C.   237 ml.

Mixing: Mix triethanolamine and denatured alcohol into mineral oil. Store in an airtight container.

Use: Apply a small amount to a soft cloth for cleaning gun parts.

CAUTIONS: Triethanolamine may be somewhat irritating to skin and mucous membranes. Denatured alcohol may be moderately toxic by ingestion and is flammable.

Notes: For proper grade of alcohol, see Appendix 4, *Denatured Alcohol.*

GUN LUBRICANT

Ingredients: Petrolatum   3/4 C.   170 g.
Light Machine Oil   1/4 C.   59 ml.

Mixing: Heat ingredients in the top of a double boiler until they can be mixed together. Cool to just above solidification point and pour into a jar.

Use: Apply a small amount to gun parts with a soft cloth.

Notes:

GUN "NITRO" SOLVENT

Ingredients: Amyl Acetate   2 T.   30 ml.
Benzene   2 T.   30 ml.
Motor Oil (SAE 30)   4 T.   60 ml.

Mixing: Mix amyl acetate and benzene into motor oil. Store in an airtight container.

Use: Swab gun barrels after firing.

CAUTIONS: Amyl acetate is toxic and benzene is flammable. Handle with care.

Notes:

GUN-POLISHING CLOTH

Ingredients: Silicone Oil Emulsion 2 T. 30 ml.
Water 1 C. 237 ml.

Mixing: Mix silicone oil emulsion into water. Saturate an 18 inch X 18 inch cloth in solution. Ring out excess liquid and dry cloth in oven.

Use: Rub cloth **slowly** over barrel, receiver, trigger, forearm, and stock. A thin film of silicone will be deposited over the surfaces, producing a soft shine and protection against rust and finger marks.

Notes:

INSECT AND MOSQUITO REPELLENT LIQUID

Ingredients: Camphor 1 T. 14 g.
Calcium Chloride 1 T. 14 g.
Isopropyl Alcohol 1/2 C. 118 ml.

Mixing: Stir camphor and calcium chloride into isopropyl alcohol until dissolved. Store in an airtight container.

Use: Use a small amount on exposed areas of skin to protect from mosquitos and other bothersome insects.

CAUTIONS: Camphor vapors are flammable. Isopropyl alcohol may be toxic by ingestion and is flammable.

Notes: For discussion on grades of alcohol, see Appendix 4, *Denatured Alcohol.*

LEATHER CLEANER I

Ingredients: Denatured Alcohol 3/4 C. 177 ml.
White Vinegar 1/2 C. 118 ml.
Water 1-1/2 C. 355 ml.

Mixing: Stir denatured alcohol and white vinegar into water. Store in an airtight container.

Use: Apply with a cloth or sponge, wipe dry, and rinse with a damp cloth or sponge.

CAUTIONS: Denatured alcohol may be toxic by ingestion and is flammable.

Notes: For proper grade of alcohol, see Appendix 4, *Denatured Alcohol.*

LEATHER CLEANER II

Ingredients:			
	Soap Powder	3 oz.	84 g.
	Household Ammonia	3 oz.	90 ml.
	Glycerin	3/4 oz.	22 ml.
	Water	5 oz.	150 ml.
	Ethylene Dichloride	1/2 oz.	14 ml.

Mixing: Mix soap powder, household ammonia, and glycerin into water, stirring thoroughly. Add ethylene dichloride. Store in an airtight container.

Use: To clean leather, apply with a sponge or cloth.

CAUTIONS: Ethylene dichloride is toxic by ingestion, inhalation, and skin absorption and is also caustic to skin. If contact is made with skin, flush freely with clear water. Do not inhale household ammonia.

Notes:

WHITE LEATHER DRESSING

Ingredients:			
	Water	7/8 C.	210 ml.
	Titanium Dioxide	5 T.	70 g.
	Stearic Acid	1/2 t.	2 g.
	Trisodium Phosphate	1 t.	4 g.

Mixing: Slowly heat water, adding remaining ingredients until dissolved. Cool and pour into container. Store in an airtight container.

Use: Apply to leather with a cloth or small sponge.

CAUTIONS: Trisodium phosphate is a skin irritant and moderately toxic by ingestion. Use rubber gloves.

Notes:

FUNGICIDE AND MILDEWPROOFER FOR LEATHER

Ingredients:			
	Copper Sulfate	1-1/2 oz.	42 g.
	Water	1 pt.	473 ml.

Mixing: Mix copper sulfate and water. Store in an airtight container.

Use: Immerse flannel cloth in solution until saturated, wring out excess, and dry flannel cloth. To use, simply rub cloth over leather to be protected.

CAUTIONS: Copper sulfate is highly toxic by ingestion. Use with caution.

Notes:

LEATHER GLOVE CLEANER

Ingredients:			
Coconut Oil	2 oz.	60 ml.
Castor Oil	1 oz.	30 ml.
Neatsfoot Oil	1 oz.	30 ml.
Olive Oil	1 oz.	30 ml.
Stoddard Solvent	1 pt.	473 ml.

Mixing: Mix coconut oil, castor oil, neatsfoot oil, and olive oil into stoddard solvent and stir thoroughly.

Use: Dip gloves repeatedly in and out of solution and rub dry.

CAUTIONS: Stoddard solvent is mildly flammable and toxic if taken internally.

Notes:

LEATHER MILDEW REMOVER

Ingredients:			
Sodium Bicarbonate (Baking Soda)	1 C.	227 g.
Water	to suit	

Mixing: Make paste of sodium bicarbonate and water.

Use: Rub into leather and let stand in sun or warm area for 24 hours. Mildew will be killed, but leather will need to be repolished to restore finish.

Notes:

LEATHER PRESERVATIVE I

Ingredients:			
Neatsfoot Oil	1-1/2 C.	355 ml.
Mineral Oil	1-1/2 C.	355 ml.

Mixing: Stir ingredients together.

Use: Apply liberally to leather with a cloth, allow to soak in for a few minutes, then rub dry with a clean cloth.

Notes: For each of these leather preservatives, penetration is improved if the mixture is warmed before applying it.

LEATHER PRESERVATIVE II

Ingredients: Neatsfoot Oil   1-1/2 C.   355 ml.
            Castor Oil   1-1/2 C.   355 ml.

Mixing: Stir ingredients together.

Use: Apply liberally to leather with a cloth, allow to soak in for a few minutes, then rub dry with a clean cloth.

Notes:

LEATHER PRESERVATIVE III

Ingredients: Oleic Acid   4 T.   60 ml.
            Triethanolamine   1 t.   5 ml.
            Neatsfoot Oil   1-3/4 C.   414 ml.
            Water   1-1/2 C.   355 ml.

Mixing: Mix oleic acid and triethanolamine with 1/2 (207 ml.) of neatsfoot oil. Slowly stir in 1/2 (177 ml.) of water. Then stir in the balance of neatsfoot oil, and finally the balance of water.

Use: Apply liberally to leather with a cloth, allow to soak in for a few minutes, then rub dry with a clean cloth.

CAUTIONS: Triethanolamine may be somewhat irritating to skin and mucous membranes.

Notes:

LEATHER WATERPROOFING

Ingredients: Silicone Oil   1 T.   15 ml.
            Stoddard Solvent   1 C.   237 ml.

Mixing: Stir silicone oil into stoddard solvent.

Use: Rub on liberally with a cloth.

CAUTIONS: Stoddard solvent is mildly flammable.

Notes:

MOISTURE PROTECTION FOR CASED GUNS OR FISHING EQUIPMENT

Ingredients: Silica Gel   1/4 cup   57 g.
Cloth Bag

Mixing: Spread silica gel on pan and heat in oven at 400°F. for 2 hours to drive out moisture. Put silica gel into cloth bag and tie closed. Place bag in gun case or fishing rod and reel case for moisture protection. Periodically, remove silica gel from bag and repeat drying procedure.

Notes:

MOSQUITO REPELLENT POWDER

Ingredients:		
Talc	4 T.	56 g.
Cornstarch	1-3/4 C.	397 g.
Eucalyptus Oil	2 T.	30 ml.

Mixing: Mix talc and cornstarch, then stir in eucalyptus oil until it is entirely absorbed.

Use: Dust on clothes, on the outside of sleeping bags, etc.

Notes:

NEATSFOOT OIL EMULSION

Neatsfoot oil is probably the best preservative and waterproofing material ever found for leather. It is made from the hooves and shinbones of cattle. But it is difficult to impregnate leather with pure neatsfoot oil. The following emulsion was developed to improve the penetrating power of the oil.

Ingredients:		
Oleic Acid	4 T.	60 ml.
Triethanolamine	1 t.	5 ml.
Neatsfoot Oil	14 oz.	420 ml.
Water	12 oz.	360 ml.

Mixing: Mix oleic acid and triethanolamine with 1/2 (210 ml.) of neatsfoot oil at room temperature. Add water gradually. Continue stirring and add balance of neatsfoot oil and water.

Use: Apply with a soft cloth, rub in well.

CAUTIONS: Triethanolamine may be somewhat irritating to skin and mucous membranes.

Notes:

OIL BOOT DUBBING

Ingredients:			
Neatsfoot Oil		5 T.	75 ml.
Tallow		1 T.	14 g.
Mineral Oil		1-1/2 T.	22 ml.

Mixing: Heat ingredients in the top of a double boiler until tallow melts and can be stirred in with the other ingredients. Cool to room temperature and bottle.

Use: Apply to boots and shoes with an old toothbrush, especially in the seams for waterproofing.

Notes:

SADDLE SOAP

Ingredients:			
Soap Powder		3/4 C.	170 g.
Water		3-1/2 C.	829 ml.
Neatsfoot Oil		1/4 C.	59 ml.
Beeswax		1/2 C.	113 g.

Mixing: Heat soap powder and water slowly until soap powder is dissolved. Separately heat neatsfoot oil and beeswax in the top of a double boiler until beeswax melts and can be mixed with neatsfoot oil. Slowly add the contents of the double boiler to the soap and water solution, stirring until the mixture thickens and cools somewhat. Then pour into containers and cool to room temperature.

Use: Apply to leather with a damp sponge, rub until a light lather forms, then wipe dry with a clean cloth.

Notes:

SKI WAX

Ingredients:			
Wood Tar	1-3/4 C.	414 ml.
Diglycol Stearate	1/2 C.	113 g.
Carnauba Wax	1/4 C.	57 g.

Mixing: Melt all ingredients together in the top of a double boiler until they can be stirred together. When partially cool, pour into molds such as small cardboard boxes the size of kitchen matchboxes.

Use: Rub well onto skis.

Notes:

TENNIS AND BADMINTON NET PRESERVATIVE

Ingredients:		
Copper Naphthenate	3/4 C.	170 g.
Fuel Oil	1/4 C.	59 ml.
Creosote	3/4 C.	177 ml.
Naphtha	3 C.	711 ml.

Mixing: Mix copper naphthenate, fuel oil, and creosote into naptha, stirring well. Store in an airtight container.

Use: Saturate nets in this solution and hang up to dry. This will protect against mildew and dry rot.

CAUTIONS: Copper naphthenate is moderately toxic by ingestion. Naphtha is highly flammable. Use rubber gloves.

Notes:

TENNIS RACKET STRING PRESERVATIVE

Ingredients:		
Water	2 qts.	2 L.
Oleic Acid	1/2 oz.	15 ml.
Glycerin	4 oz.	120 ml.

Mixing: Heat water to boiling and add oleic acid and glycerin with stirring. Cool down to room temperature.

Use: Apply to strings of racket with a brush, allow to dry, and wipe with a soft, dry cloth.

Notes:

WATERPROOFING MATCHES

Ingredients: Paraffin Wax   Small Quantity
              Large-size Kitchen Matches   6 or 8
              Suitable container—Any small box will do.

Mixing: Heat paraffin wax in top of a double boiler until melted. Dip heads of matches into melted paraffin wax, allowing wax to cool and solidify on matchhead before dipping for second or more coats.

Use: Always carry treated matches in outdoor clothing, thus assuring that a fire can be built under wet conditions.

Notes:

"We forgot the food!"

pH Preferences
of Some Plants

In light of upward spiraling prices and a world food shortage, growing our own food is becoming a way of life for many of us. And it's fun. But it's work too. Ask any weekend gardener who discovers new muscles on Monday morning. So it just makes sense to take advantage of any practice that will increase yields for the same amount of effort. And one of the most rewarding things that can be easily done is to provide the plant with soil that has been adjusted to the most desirable degree of acidity or alkalinity. This is known as adjusting the pH of the soil.

For practical purposes the range of pH we are concerned with is from 4 to 8, with 7 representing neutral. As you go down the scale from 7, the acidity increases. As you go up the scale from 7, the alkalinity increases. The procedures for testing the pH of soil are simple, and will be covered in another paragraph.

Now let's go through a typical example of adjusting the soil in a given area to make it ideal for the plants we wish to grow there. By reference to the following table, we find that asparagus, beets, cabbage, carrots, cauliflower, celery, lettuce, onions and parsley all have a pH preference range of 7 to 8 with the ideal being 7.5 on the scale, which is slightly alkaline. So naturally it follows that these vegetables should be planted in the same area where the soil has been adjusted to a pH of approximately 7.5. Of course, the same procedure applies to other areas where plants of other pH preferences will be grown. In the case of house plants, each pot of soil will be individually adjusted to the plant's preference.

To determine and alter soil pH, you will need an inexpensive tester. There are a number of these on the market that may be obtained from garden supply centers and swimming pool supply dealers. The writer's preference is to use chemically treated paper that changes color to indicate pH. This paper comes in rolls about 1/4" wide housed in a plastic tape-type dispenser. The procedure we use is as follows.

Step 1) Let's assume you have set aside an area 5' wide by 20' in length to grow the group of vegetables that prefer a pH of 7.5. Your objective is to adjust the soil in that area to a pH of approximately 7.5. First, get a glass container that will hold at least a pint of water. (A Pyrex measuring cup works fine.) You will also need either a plastic or wooden spoon. Rinse the inside of the glass container and the spoon with *distilled* or *demineralized* water. These are commonly used for steam irons, and are available at any supermarket. *Be sure your hands do not touch the inside of the glass or bowl of the spoon,* as this could contaminate your reading.

Step 2) To obtain an average sample of the area in which you want to determine and adjust the pH, place a spoonful of soil taken from locations in the area about 3' apart, and several inches below the surface. Mix these samples together in the glass, thoroughly. Next, cover the soil with distilled or demineralized water to about 1/2" over the soil. Now, using your spoon, mix the water and soil thoroughly and allow glass container to stand undisturbed. In a short time, the soil will settle to the bottom of the container leaving the clear water over it.

Step 3) You are now ready to determine what the pH of the area is. Pull out about a 2" strip of the chemically treated paper from the dispenser and hold by one end. Do not allow your fingers to come into contact with the rest of the paper strip. Immerse one end of the paper in the clear water over the soil and compare the wet section of the strip with the color chart found on the side of the container. The match of color between the strip and chart will tell you the pH of the soil.

Step 4) Formulas for adjusting the pH of house plants are not generally economical for larger areas such as gardens. For this use, the most effective and economical method is as follows.

To alkalize the soil, i.e., to raise the pH number, the least expensive and best material to use is ground limestone. The formula is 6 pounds of limestone for each 100 square feet of area to increase the pH by one point. Therefore, if the pH of your soil is 6.5 and the desired level is 7.0, you would need to work in 3 pounds of limestone for the 100 square foot area.

On the other hand, if the soil needed to be acidified, the formula would be to use either aluminum sulphate or powdered sulphur, whichever is lower in cost and most readily available in your area. Applied at the rate of 2 pounds for each 100 square feet of area, the pH will be lowered by one point on the scale. Thus, if your soil had a pH of 8 and you wanted 7.5, you would have to work in 1 pound of either material over the 100 square foot area.

We are constantly adjusting soil pH on our research and testing facility near Tucson, Arizona, and find two methods of application to be effective. Either a small seed spreader with an adjustable discharge slot, or a rotary seeder or duster, also adjustable, will do a good job of giving even distribution over the area. After the material has been distributed, we work it into the soil with a rototiller. Of course for small areas, an ordinary garden rake will do just as well.

It is important to remember that some time is required for the full acidifying or alkalizing process to complete itself. Therefore, the most desirable time to test and adjust is in the fall, and then make a final check before planting time.

Following is a list of vegetables, fruits, flowers and grasses, showing their acceptable pH range. The optimum value is at the midpoint of the range. For example, the acceptable range for asparagus is 7-8, so the optimum value is 7.5.

VEGETABLES AND FRUITS

Name	pH Range	pH Optimum
Asparagus	7–8	7.5
Beets	7–8	7.5
Cabbage	7–8	7.5
Carrots	7–8	7.5
Cauliflower	7–8	7.5
Celery	7–8	7.5
Lettuce	7–8	7.5
Onions	7–8	7.5
Parsley	7–8	7.5
Plums	7–8	7.5
Broccoli	6–7	6.5
Brussels Sprouts	6–7	6.5
Corn	6–7	6.5
Cucumbers	6–7	6.5
Peas	6–7	6.5
Peppers	6–7	6.5
Radishes	6–7	6.5
Raspberries	6–7	6.5
Rhubarb	6–7	6.5
Spinach	6–7	6.5
Melons	6–7	6.5
Beans	5–6	5.5
Citrus	5–6	5.5
Parsnips	5–6	5.5
Potatoes	5–6	5.5
Grapes	5–6	5.5
Squash	5–6	5.5
Strawberries	5–6	5.5
Tomatoes	5–6	5.5
Turnips	5–6	5.5

FLOWERS

Name	pH Range	pH Optimum
Barberry	7–8	7.5
Calendula	7–8	7.5
Geranium	7–8	7.5
Morning Glory	7–8	7.5
Nasturtium	7–8	7.5
Petunia	7–8	7.5
Poppy	7–8	7.5
Sweet Pea	7–8	7.5

Alyssum	6–7	6.5
Aster	6–7	6.5
Candytuft	6–7	6.5
Cauna	6–7	6.5
Carnation	6–7	6.5
Chrysanthemum	6–7	6.5
Columbine	6–7	6.5
Cosmos	6–7	6.5
Crocus	6–7	6.5
Dahlia	6–7	6.5
Dogwood	6–7	6.5
Feverfew	6–7	6.5
Gladiolus	6–7	6.5
Hollyhock	6–7	6.5
Hyacinth	6–7	6.5
Hydrangia (Pink)	6–7	6.5
Iris	6–7	6.5
Marigold	6–7	6.5
Pansy	6–7	6.5
Peony	6–7	6.5
Rose	6–7	6.5
Snapdragon	6–7	6.5
Tulip	6–7	6.5
Violet	6–7	6.5
Zinnia	6–7	6.5
Delphinium	5–6	5.5
Easter Lily	5–6	5.5
Fern	5–6	5.5
Lupine	5–6	5.5
Begonia	5–6	5.5
Phlox	5–6	5.5
Primrose	5–6	5.5
Azalea	4–5	4.5
Holly	4–5	4.5
Hydrangia (Blue)	4–5	4.5
Rhododendron	4–5	4.5

GRASSES

Blue Grass	7–8	7.5
Clover	7–8	7.5
Squirrel Tail Grass	7–8	7.5
Red Clover	6–7	6.5
Bermuda Grass	6–7	6.5
Colonial Bent Grass	6–7	6.5
Creeping Bent Grass	6–7	6.5
Italian Rye Grass	6–7	6.5

Perennial Rye Grass	6–7	6.5
Rough Blue Grass	6–7	6.5
Sudan Grass	5–6	5.5
Panic Grass	4–5	4.5

TYPES OF MIXING

1. Dry materials that are simply mixed together. Example:
   Fire Extinguishing Powder
   Mix fine silica mason sand and sodium bicarbonate.
   These are both simply mixed together.

2. Dry or semisolid materials that become liquids when combined with a solvent such as water or alcohol. Example:
   Cuticle Remover
   Stir trisodium phosphate (TSP) into glycerin mixing thoroughly, and add water to form a paste. In this type of product we have put a powder into a solution in a solvent, and our end product becomes a liquid.

3. Materials that require heat to turn them from a solid or semisolid into a liquid and back into a solid on cooling. Example:
   Waterproofing Matches
   Melt paraffin wax in double boiler just above solidification point. Dip individual matches into the molten wax to about half the length. Pull out and allow the wax to solidify. In this type of compound we have converted a solid and semisolid into a liquid by melting it, and returned it to a semisolid by cooling.

4. Emulsions. Example:
   Astringent Skin Cream
   Mix mineral oil and beeswax together until the beeswax is melted and mixed with the mineral oil. In separate pan, heat water and stir in borax and powdered alum until dissolved. Pour this mixture slowly into mineral oil and beeswax. When cool, just above the solidification point, add oil-soluble perfume. As we all know, water and oil will not mix unless the oil is combined with an emulsifying agent.

SOME SCIENTIFIC PRINCIPLES

ABRASION (Abrasives): All surfaces, including wood, metal, plastic, glass, etc., no matter how smooth they appear to be, consist of ridges and serrations. The act of polishing consists of abrading these surfaces with an abrasive that is harder than the surface being polished. (See Metal Polish.)

ABSORPTION/ADSORPTION: These two are frequently confused.

ABSORPTION: This is the attraction of one substance by another so that the absorbed substance disappears physically and becomes an integral part of the absorbing substance. Here are some examples. Canvas Fireproofing Compound—In this formula the ammonium phosphate and ammonium chloride are absorbed into the water. Laundry Water Softener—Here the sodium silicate (waterglass) and the soda ash (sodium carbonate) are absorbed by the washing water. And there are many other examples of course, such as sugar in coffee or tea, and salt in water.

ADSORPTION: This is the adhering of atoms, ions or molecules of one substance to another substance which is called the adsorbent. An adsorbent such as silica gel, activated carbon, or activated alumina is a material that has a vast internal surface area because of its porosity. Unlike an absorbent, such as sugar dissolved in water, an adsorbent does not change in its physical form, and may be reactivated by heat and used over and over.

EMULSIONS: Emulsions are a stable mixture of two or more immiscible liquids that are held in suspension by small amounts of chemicals called emulsifiers. A mixture of oil and water is a good example. Emulsifiers are of two basic types: (1) polymers (complex molecules made by catalyzing two simple molecules) of protein and carbohydrates act to coat the oil particles which prevents them from adhering to each other to form a mass, and (2) fatty acids that act to reduce surface tension so the droplets remain separated.

Good examples of both types of emulsions are found in the formulas for Mineral Oil Emulsion, Soap, Basic and Rug and Carpet Cleaner.

STATE OF MATTER: (Solid, Liquid, Gas) An excellent example of the changes in matter from solid to liquid to gas and back to solid is the candle.

Before the wick is lit, the candle is, of course, a solid. But when the wick is lit, the heat of the flame melts the wax around it, converting it to a liquid. As the liquid wax is drawn up the wick to its tip by **capillary action** (much as ink migrates in a blotter), it converts to a gas in which form it is burned. Then, if the flame is snuffed out, the process reverses itself and the section of the candle which was previously a liquid turns back into a solid.

Here's another common demonstration of the change in matter from a solid to a liquid to a gas. When water is frozen to ice, it becomes a solid. When it attracts heat from the atmosphere around it, it turns to a liquid. While in its liquid form, it vaporizes into a gas which passes off to the surrounding atmosphere.

DENATURED ALCOHOL

ETHYL ALCOHOL/DENATURED ETHYL ALCOHOL

Alcohols are widely used in many areas of chemistry, and especially in compounding of formulas such as those found in *The Formula Manual.* As a matter of fact, it would be hard to conceive of being able to make many of the compounds without it. But, based on the mail we have received from many teachers and students, there seems to be some confusion over the two primary types, i.e., ethyl alcohol in its pure form, and ethyl alcohol that has been denatured, denatured alcohol. We hope this section will promote a better understanding for those who may not be completely clear on the subject.

Ethyl alcohol (ethanol, grain alcohol) (C_2H_2OH or CH_3CH_2OH), is a clear colorless liquid having a melting point of -117°C., and a boiling point of 78.5°C. It is miscible in any proportion with water or ether, and is soluble in a sodium hydroxide (caustic soda) solution. Flammable, it burns in air with a bluish transparent flame, producing water and carbon dioxide as it burns. Density is 0.789 at 20°C.

Absolute (anhydrous) ethyl alcohol is obtained by the removal of water. One process for accomplishing this is to react the water in the alcohol, with calcium oxide and then distill the alcohol.

Ethyl alcohol is made by (A) the fermentation of grains and fruits, and also directly from dextrose, (B) by absorption of ethylene from coal or petroleum gas, and then water reaction, and (C) by the reduction of acetaldehyde in the presence of a catalyst.

Ethyl alcohol is used in tremendous quantities in beverages which are taxed by the federal government. There are many other uses as well, such as in pharmaceuticals, tinctures, and extracts for internal use, where it is not taxed as it is for beverages. For these uses however, a special tax-free permit must be obtained from the Alcohol and Tobacco Unit of the Federal government. Permits of this type are available to educational institutions as well. However, for small quantity use, just buying a bottle of 95 or 100 proof vodka is much less complicated.

Denatured alcohol is ethyl alcohol (the same as is used in beverages), except that it has been adulterated with other chemicals that make it unfit for beverage use, while still retaining its other characteristics. Therefore, denatured alcohol is not taxed as pure ethyl alcohol is, making it very much less expensive.

There are two basic types of denatured alcohol, Completely Denatured (CDA), and Specially Denatured (SDA). The denaturants that are used are specified by the Alcohol and Tobacco Tax Unit, and depend on the end use of the alcohol. For example. a denaturant acceptable for use in alcohol to be used as an industrial solvent would be entirely unacceptable for use in a body lotion or mouthwash because of its degree of toxicity and irritating properties. Therefore, the type of denatured alcohol must be chosen for the compound it's to be used in. Following is a list of general compound classifications and the code number of the denatured alcohol approved for each. You will notice that a given type of denaturant may be used in the alcohol that is used in many different formulas. From the following chart it will be seen that Specially Denatured Alcohol, Type 40, is approved for use in a number of applications such as: bath preparations, bay rum, cleaning solutions, colognes, etc. Therefore, in purchasing denatured alcohol it is practical to select the type that fits as many uses as possible. While the approved types of denatured alcohol for specific uses is mandatory for a manufacturer who resells, it does not apply to the individual making the product for his own use. However, in the interest of safety, it is *highly recommended* that only approved types for the specific formulas be used.

DENATURED ALCOHOL USE	ALCOHOL and TOBACCO TAX APPROVED TYPE
Animal Feed Supplements	35A.
Antifreeze	1.
Antiseptic Bathing Solutions	46.
Antiseptic Solutions	23A, 37, 38B, 38F.
Bath Preparations	1, 3A, 3B, 23A, 30, 36, 38B, 39B, 39C, 40, 40A, 40B, 40C.
Bay Rum	23A, 37, 38B, 39, 39B, 39D, 40, 40A, 40B, 40C.
Brake Fluids	1, 3A.
Candy Glazes	13A, 23A, 35, 35A.
Cellulose Coatings	1, 23A, 30.
Cleaning Solutions	1, 3A, 23A, 23H, 30, 36, 39B, 40, 40A, 40B, 40C.
Coatings	1, 23A.
Colognes	38B, 39, 39A, 39B, 39C, 40, 40A, 40B, 40C.
Cutting Oils	1, 3A.
Dentifrices	31A, 37, 38B, 38C, 38D.
Deodorants (Body)	23A, 38B, 39B, 39C, 40, 40A, 40B, 40C.

Detergents (Home Use)	1, 3A, 23A, 23H, 30, 36, 39B, 40, 40A, 40B, 40C.
Detergents (Industrial)	1, 3A, 23A, 30.
Disinfectants	1, 3A, 3B, 23A, 23H, 27A, 27B, 30, 37, 38B, 39B, 40, 40A, 40B, 40C.
Drugs and Medicinal Chemicals	1, 2B, 2C, 3A, 6B, 12A, 13A, 17, 29, 30, 32.
Duplicating Fluids	1, 3A, 30.
Dye Solutions	1, 3A, 23A, 30, 39C, 40, 40A, 40B, 40C.
Fuel Uses	1, 3A, 28A.
Fungicides	1, 3A, 3B, 23A, 23H, 27A, 27B, 30, 37, 38B, 39B, 40, 40A, 40B, 40C.
Hair and Scalp Preparations	3B, 23A, 23F, 23H, 37, 38B, 39, 39A, 39B, 39C, 39D, 40, 40A, 40B, 40C.
Inks	1, 3A, 13A, 23A, 30, 32, 33.
Insecticides	1, 3A, 3B, 23A, 23H, 27A, 27B, 30, 37, 38B, 39B, 40, 40A, 40B, 40C.
Iodine Solutions and Tinctures	25, 25A.
Lacquer Thinners	1, 23A.
Liniments	27, 27B, 38B.
Lotions and Creams (Body, Face and Hands)	23A, 23H, 31A, 37, 38B, 39, 39B, 39C, 40, 40A, 40B, 40C.
Mouthwashes	37, 38B, 38C, 38D, 38F.
Perfumes and Tinctures	38B, 39, 39B, 39C, 40, 40A, 40B, 40C.
Petroleum Products	1, 2B, 3A.
Plastics-Cellulose	1, 2B, 3A, 12A, 13A, 30.
Plastics and Resins	1, 2B, 3A, 12A, 13A, 30.
Polishes	1, 3A, 30, 40, 40A, 40B, 40C.
Preserving Solutions	1, 3A, 12A, 13A, 22, 23A, 30, 32, 37, 38B, 42, 44.
Resin Coating (Natural)	1, 23A.
Resin Coating (Synthetic)	1, 23A, 30.
Room Deodorants	3A, 22, 37, 38B, 39B, 39C, 40, 40A, 40B, 40C.
Rubbing Alcohol	23H.
Scientific Instruments	1, 3A.
Shampoos	1, 3A, 3B, 23A, 27B, 31A, 36, 38B, 39A 39B, 40, 40A, 40B, 40C.
Shellac Coatings	1, 23A.

Soaps (Industrial)	1, 3A, 23A, 30.
Soaps (Toilet)	1, 3A, 3B, 23A, 30, 36, 38B, 39B, 39C, 40, 40A, 40B, 40C.
Soldering Flux	1, 3A, 23A, 30.
Solutions (Miscellaneous)	1, 3A, 23A, 30, 39B, 40, 40A, 40B, 40C.
Solvents and Thinners	1, 23A.
Stains (Wood)	1, 3A, 23A, 30.
Sterilizing Solutions	1, 3A, 12A, 13A, 22, 23A, 30, 32, 37, 38B, 42, 44.
Toilet Water	38B, 39, 39A, 39B, 39C, 40, 40A, 40B, 40C.
Unclassified Uses	1, 3A.
Vinegar	18, 29, 35A.

DENATURING FORMULAS

While the formulas for denaturing ethyl alcohol for various applications are not particularly relevant to the user of *The Formula Manual* because the alcohol you purchase will already be denatured, they are of general interest in that they show the different degrees of contamination required for uses in different product formulas. For this reason they are included in this section.

ALCOHOL & TOBACCO TAX,
APPROVED TYPES   DENATURING FORMULAS

1)   100 gallons ethyl alcohol, 5 gallons wood alcohol.

2B)   100 gallons ethyl alcohol, 5 gallons methyl alcohol.

2C)   100 gallons ethyl alcohol, 33 pounds metallic sodium and 1/2 gallon benzene.

3A)   100 gallons ethyl alcohol, 5 gallons methyl alcohol.

3B)   100 gallons ethyl alcohol, 1 gallon pine tar.

6B)   100 gallons ethyl alcohol, 1/2 gallon pyridine bases.

12A)   100 gallons ethyl alcohol, 5 gallons benzene.

13A)   100 gallons ethyl alcohol, 10 gallons ethyl ether.

17)   100 gallons ethyl alcohol, 64 fluid ounces bone oil.

18)   100 gallons ethyl alcohol, 100 gallons vinegar (90Gr.)

23A)   100 gallons ethyl alcohol, 10 gallons acetone.

23F)   100 gallons ethyl alcohol, 3 pounds salicylic acid, USP, 1 pound resorcin, USP, 1 gallon bay oil, USP.

23H)   100 gallons ethyl alcohol, 8 gallons acetone, 1.5 gallons methyl isobutyl ketone.

27A)   100 gallons ethyl alcohol, 35 pounds camphor, USP, 1 gallon clove oil, USP.

27B) 100 gallons ethyl alcohol, 1 gallon lavender oil, USP, 100 pounds medicinal soft soap, USP.

28A) 100 gallons ethyl alcohol, 1 gallon gasoline.

29)   100 gallons ethyl alcohol, 1 gallon 100% acetaldehyde.

30)   100 gallons ethyl alcohol, 10 gallons methyl alcohol.

31A) 100 gallons ethyl alcohol, 100 pounds glycerol, USP, 20 pounds hard soap.

32)   100 gallons ethyl alcohol, 5 gallons ethyl ether.

37)   100 gallons ethyl alcohol, 45 fluid ounces eucalyptol USP, 30 ounces by weight thymol, 20 ounces by weight menthol USP.

38B) 100 gallons ethyl alcohol, 10 pounds menthol, USP.

38C) 100 gallons ethyl alcohol, 10 pounds menthol, USP, 1.25 gallons formaldehyde, USP.

38F) 100 gallons ethyl alcohol, 6 pounds boric acid, USP, 1-1/3 pounds thymol, 1-1/3 pounds chlorothymol, and 1-1/3 pounds menthol, USP.

39)   100 gallons ethyl alcohol, 9 pounds sodium salicylate USP, 1.25 gallons extract of quassia, 1/8 gallon tert.-butyl alcohol.

39A) 100 gallons ethyl alcohol, 60 ounces quinine, 1/8 gallon tert.-butyl alcohol.

39B) 100 gallons ethyl alcohol, 2-1/2 gallons diethyl phthalate, 1/8 gallon tert.-butyl alcohol.

39C) 100 gallons ethyl alcohol, 1 gallon diethyl phthalate.

39D) 100 gallons ethyl alcohol, 1 gallon bay oil, 50 ounces by weight quinine sulphate.

40)   100 gallons ethyl alcohol, 1-1/2 ounces brucine, 1/8 gallon tert.-butyl alcohol.

40A) 100 gallons ethyl alcohol, 1 pound sucrose octaacetate, 1/8 gallon tert.-butyl alcohol.

40B) 100 gallons ethyl alcohol, 1/16 ounce denatonium benzonate, 1/8 gallon tert.-butyl alcohol.

40C) 100 gallons ethyl alcohol, 3 gallons tert.-butyl alcohol.

42)   100 gallons ethyl alcohol, 80 grams potassium iodine, USP, 109 grams red mecuric iodide.

44)   100 gallons ethyl alcohol, 10 gallons n-butyl alcohol.

GRADES OF MATERIAL

The selection of materials depends on the end use of the product. For example, the selection of denatured alcohol for use in a paint or varnish thinner would be entirely different from the choice of a product that would be in contact with the skin. (See Appendix 4.)

There are two basic grades of chemicals used: 1) U.S.P., and 2) Manufacturing or Technical.

1. *U.S.P.,* is an abbreviation for the United States Pharmacopoeia which is the official federal book of chemicals and drugs. This publication sets up the standards of purity and other specifications that the manufacturer must comply with. Generally speaking, U.S.P. grades are used in compounds that are taken internally, or come in contact with delicate areas of the body that require pure materials. An example of this is where a material such as magnesium sulfate (epsom salts) is contained in a product for internal use, it must be U.S.P. grade.

2. *Manufacturing or Technical Grade.*
In this category, the standards for purity are understandably less than in the U.S.P. grade, in that the end product does not directly affect human health. For example, if the epsom salts were to be used in a foot bath, the purity requirements would not be the same as for internal use, and the Manufacturing or Technical grade would be acceptable.

There is a substantial difference in the cost of these two grades of materials, therefore, the selection should always be made on the basis of the end use.

The *odor and color,* used in a preparation is largely a matter of personal choice, and usually has no effect on the function of the compound itself. For example, in the formula for Face Lotion, the perfume in the product has no effect on the properties it imparts to the skin. However, if perfume or color is desired, it must be of a type that is compatible with the compound it is to be used in.

Dyes and perfumes fall into three general categories: (1) those that are soluble in oil, (2) those that are soluble in water or alcohol, and (3) those that are suitable for use in an emulsion.

Examples of these types are as follows:

Type One; is soluble in oil and would be used in liquids, semisolids, and solids having an oil base such as Baby Oil.

Type Two; is soluble in water and alcohol and would therefore be used in compounds such as Face Lotion where the base is water and alcohol.

Type Three; is an emulsion type and logically, is used in emulsions. Each formula that requires a perfume or dye has the type specified in the formula itself.

But a word of warning. Perfumes and dyes are in highly concentrated form, and should be used very sparingly.

FORMULATING TIPS AND PROCEDURES

The ingredients in each formula must be combined in the correct sequence, because a chemical reaction may take place and a deviation from that sequence could prevent it. One of the best ways to insure against error is to use separate containers for the ingredients, numbered in the order that they are to be incorporated into the compound. Paper cups work well (See Figure 1.) for this, in that they are inexpensive, disposable, and can be easily numbered with a felt marker.

The correct measurement of ingredients is important. Always have the utensil that is being measured into on a level surface. (See Figure 2.) Measure accurately to the line of the quantity specified. Avoid touching container to eliminate the possibility of compacting the material, which would increase its quantity. (See Figure 11.)

Use standard measuring spoons, not eating spoons. A spoonful is a measuring spoon that is filled level with the top. Dip spoon in material to rounding, and then scrape excess off with a knife. (Figure 11a.)

A speck is the amount of powdered or granular material that will lie in a 1/4" square marked on a piece of paper. (See Figure 12a.)

When a formula calls for a portion of a cup of lump or semisolid material that is *not water soluble,* use this method. If you want to measure 1/2 cup of lump paraffin for example, pour *cold* water in a measuring cup to the 1/2 cup line. Next, add lumps of paraffin to the water until the water level reaches the 1 cup line. Pour out the water, and the paraffin in the cup will equal 1/2 cup. (See Figures 12 and 13.)

Follow instructions given in each individual formula, and *always* follow the labeling instructions which follow.

All chemicals that are stored in containers should be labeled, regardless of whether they are a raw material or a finished compound. This is basic, and must be followed in the interest of safety. Keep all chemicals out of the reach of children, and note the contents on the label. In this way if a child, or even an unsuspecting adult, should accidentally consume the contents, the doctor would know what treatment to initiate. While these formulas have been chosen with an eye to safety, many materials normally regarded as safe can be dangerous if taken internally, or to excess.

We recommend that those chemicals or formulas that are toxic be clearly labeled as "poison" and marked with a skull and crossbones or some other widely recognized warning. Here are two examples of safe labels.

```
This Bottle Contains _____
Its Ingredient(s) are: _____

CAUTIONS:_____
Keep out of reach of children.

Made by _____

Date _____
              KEEP BOTTLE SEALED
```

```
This Bottle Contains _____
Its Ingredient(s) are: _____
                                      POISON
CAUTIONS:                                ☠
Keep out of reach of children.

Made by _____

Date _____
              KEEP BOTTLE SEALED
```

We have written this book with an eye to simplicity, and most of the equipment you will need can be found in most kitchens. Here is a list of the basics, and some illustrations of them.

1. Several glass measuring cups. (See Figure 2.)

2. A set of mixing bowls made of glass, ceramic or plastic.(See Figure 10.)

3. A wood fork with spacing of about 1/8" between tines. (See Figure 3.)

4. An egg beater. (See Figure 4.) An electric mixer, with beaters and bowl, is helpful but not essential. If it has variable speeds, it can be used for both wet and dry mixing, saving a great deal of time and assuring a "good blend."

5. A stem type thermometer is convenient, but again, not absolutely essential. (See Figure 5.) If one is not available, remember that water gives off a mild vapor at 140° F., a moderate vapor at 160° F., a heavy vapor at 180° F. and heavy steam at the boiling point.

6. A supply of wood tongue depressors. (See Figure 8.) These are smooth, cheap and readily available from any druggist. They make excellent mixing sticks, and are inexpensive enough to be disposable, eliminating a lot of "cleaning up."

7. Paper cups are ideal for small batch formulating. They are inexpensive, disposable and can be easily numbered or marked with a felt marker.

8. Double boilers are required in many instances. These should be Pyrex. (See Figure 6.)

9. A rubber syringe for measuring out drops. (See Figure 7.)

10. A set of standard measuring spoons. (See Figure 11a.)

11. A plastic cone and filter paper, such as is used in coffee making. (See Figure 9.) While filtering a liquid compound after it is finished is not usually essential, it is always desirable, in that a clearer, better looking product results.

12. Containers for the finished product are a matter of personal preference. In most homes, jars and bottles are available. If they are to be purchased, many supermarkets carry them, and drugstores have them for their own use. Larger quantities can be had from bottle distributors, listed in the Yellow Pages.

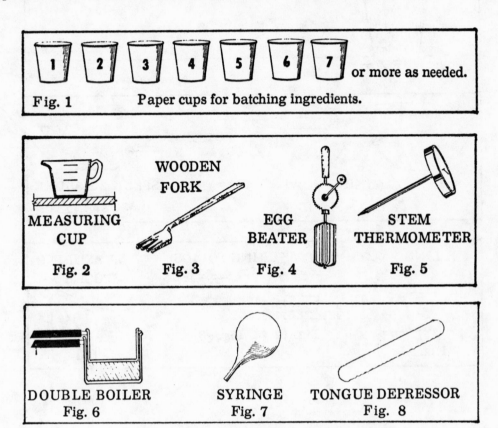

Fig. 1   Paper cups for batching ingredients.   or more as needed.

MEASURING CUP   Fig. 2

WOODEN FORK   Fig. 3

EGG BEATER   Fig. 4

STEM THERMOMETER   Fig. 5

DOUBLE BOILER   Fig. 6

SYRINGE   Fig. 7

TONGUE DEPRESSOR   Fig. 8

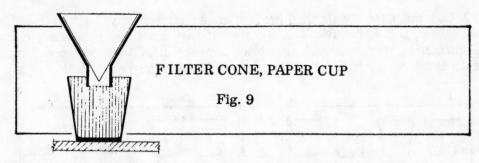

FILTER CONE, PAPER CUP

Fig. 9

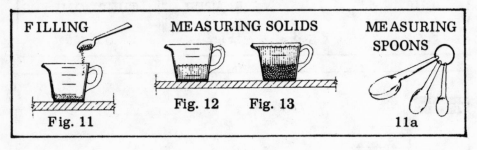

MIXING BOWLS   SPECK MEASURE

Fig. 10   Fig. 12a

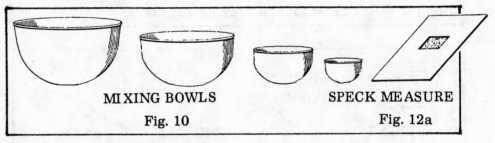

FILLING   MEASURING SOLIDS   MEASURING
                                    SPOONS

            Fig. 12   Fig. 13

Fig. 11   11a

This Bottle Contains _____
Its Ingredient(s) are:

CAUTIONS: _____

Keep out of reach of children.

Made by _____

Date _____
KEEP BOTTLE SEALED

This Bottle Contains _____
Its Ingredient(s) are:

CAUTIONS: **POISON**

Keep out of reach of children.

Made by _____

Date _____
KEEP BOTTLE SEALED

This Bottle Contains _____
Its Ingredient(s) are:

CAUTIONS: _____

Keep out of reach of children.

Made by _____

Date _____
KEEP BOTTLE SEALED

This Bottle Contains _____
Its Ingredient(s) are:

CAUTIONS: **POISON**

Keep out of reach of children.

Made by _____

Date _____
KEEP BOTTLE SEALED

This Bottle Contains _____
Its Ingredient(s) are:

CAUTIONS: _____

Keep out of reach of children.

Made by _____

Date _____
KEEP BOTTLE SEALED

This Bottle Contains _____
Its Ingredient(s) are:

CAUTIONS: **POISON**

Keep out of reach of children.

Made by _____

Date _____
KEEP BOTTLE SEALED

This Bottle Contains _____
Its Ingredient(s) are:

CAUTIONS: _____

Keep out of reach of children.

Made by _____

Date _____
KEEP BOTTLE SEALED

This Bottle Contains _____
Its Ingredient(s) are:

CAUTIONS: **POISON**

Keep out of reach of children.

Made by _____

Date _____
KEEP BOTTLE SEALED

This Bottle Contains _____
Its Ingredient(s) are:

CAUTIONS: _____

Keep out of reach of children.

Made by _____

Date _____
KEEP BOTTLE SEALED

This Bottle Contains _____
Its Ingredient(s) are:

CAUTIONS: **POISON**

Keep out of reach of children.

Made by _____

Date _____
KEEP BOTTLE SEALED

MEASUREMENT CONVERSION TABLES

3 teaspoons	equals	1 tablespoon
2 tablespoons	equals	1 liquid ounce
4 tablespoons	equals	1/4 cup
16 tablespoons	equals	1 cup
2 cups	equals	1 pint
2 pints	equals	1 quart
4 quarts	equals	1 gallon
16 ounces	equals	1 pound

METRIC CONVERSIONS

We are approaching the time when the metric system will phase out our conventional system of weights and measures. But this is a confusing transition to make, so many people are wisely beginning to learn it now. To aid in their effort, the following tables are included, and the proportions listed in each formula are expressed in both systems. Thus, by association, learning the equivalents is far easier. For simplification of measurement the metric has been rounded out to 1 decimal and specific gravity average.

CONVERSION FORMULAS

Gallons into Pounds—Multiply 8.33 (wt. 1 gallon of water) by the specific gravity (sg) and the result by the number of gallons. (See any chemical dictionary for the sg of a particular chemical.)

Pounds into Gallons—Multiply 8.33 by the sg and divide the number of pounds by the result.

Milliliters into Grams—Multiply the number of milliliters by the sg.

Grams into Milliliters—Divide the number of grams by the sg.

Milliliters into Pounds—Multiply the number of milliliters by the sg, and divide the product by 453.56 (no. of g. per lb.).

Pounds into Milliliters—Multiply the number of pounds by 453.56 and divide the product by the sg.

Milliliters into Ounces—Multiply the number of milliliters by the sg, and divide the product by 28.35 (no. g. per oz.).

Ounces into Milliliters—Multiply the number of ounces by 28.35 and divide the product by the sg.

CONVERSION FACTORS

Liquid

From	To	Multiply By
Ounces	Milliliters	29.56
Pints	Liters	0.47
Quarts	Liters	0.95
Gallons	Liters	3.78
Milliliters	Ounces	0.03
Liters	Pints	2.10
Liters	Quarts	1.05
Liters	Gallons	0.26

Dry

From	To	Multiply By
Ounces	Grams	28.35
Pounds	Kilograms	0.45
Grams	Ounces	0.035
Kilograms	Pounds	2.21

FLUID MEASURE

Metric	U.S. Regular
1 Milliliter	0.034 ounce
1 liter	33.81 ounces
1 liter	2.10 pints
1 liter	1.05 quarts
1 liter	0.26 gallons

DRY MEASURE

Metric	U.S. Regular
1 gram	0.035 ounce
1 kilogram	35.27 ounces
1 kilogram	2.21 pounds

DRY MEASURE

U.S. Regular	Metric Equivalent
1/8 teaspoon	0.54 grams
1/4 teaspoon	1.09 grams
1/2 teaspoon	2.19 grams
3/4 teaspoon	3.28 grams
1 teaspoon	4.38 grams

DRY MEASURE

U.S. Regular	Metric Equivalent
1/8 tablespoon	1.77 grams
1/4 tablespoon	3.54 grams
1/2 tablespoon	7.09 grams
3/4 tablespoon	10.63 grams
1 tablespoon	14.18 grams
1/8 ounce	3.59 grams
1/4 ounce	7.39 grams
1/2 ounce	14.18 grams
3/4 ounce	21.34 grams
1 ounce	28.35 grams
1/8 pound	56.69 grams
1/4 pound	113.39 grams
1/2 pound	226.78 grams
3/4 pound	340.17 grams
1 pound	453.56 grams
1/8 cup	28.34 grams
1/4 cup	56.69 grams
1/2 cup	113.39 grams
3/4 cup	170.08 grams
1 cup	226.78 grams

LIQUID MEASURE

U.S. Regular	Metric Equivalent
1/8 teaspoon	0.61 milliliters
1/4 teaspoon	1.23 milliliters
1/2 teaspoon	2.47 milliliters
3/4 teaspoon	3.70 milliliters
1 teaspoon	4.94 milliliters
1/8 tablespoon	1.84 milliliters
1/4 tablespoon	3.69 milliliters
1/2 tablespoon	7.39 milliliters
3/4 tablespoon	11.08 milliliters
1 tablespoon	14.78 milliliters
1/8 ounce	3.69 milliliters
1/4 ounce	7.39 milliliters
1/2 ounce	14.78 milliliters
3/4 ounce	22.17 milliliters
1 ounce	29.57 milliliters

LIQUID MEASURE

U.S. Regular	Metric Equivalent
1/8 cup	29.57 milliliters
1/4 cup	59.14 milliliters
1/2 cup	118.28 milliliters
3/4 cup	177.42 milliliters
1 cup	236.56 milliliters
1 pint	473.00 milliliters
1 quart	946.00 milliliters
1/2 gallon	1.89 liters
3/4 gallon	2.83 liters
1 gallon	3.78 liters

SOURCES OF SUPPLY AND
INDEX OF CHEMICALS

SPECIALIZED SOURCES OF CHEMICALS

The majority of ingredients in the formulas of this book are easily obtained from local sources in most areas of the United States. Borax, mineral oil, kerosene, beeswax, lanolin, paraffin wax, salt, soda ash, baking soda, and glycerin are repeatedly called for in formulating the products; most of these ingredients are in the local drug, grocery, paint, or fuel supply stores. Most of the others can be obtained from specialized stores, manufacturers, or chemical supply houses.

While it is suggested that chemicals be purchased from local sources wherever possible, some materials are not as common as others and are more difficult to locate, especially in nonmetropolitan areas. If this problem occurs, or if you prefer to order by mail at prices usually lower than those in retail stores, there are two alternatives: you can (1) buy your materials by mail from a chemical repackager, who buys chemicals in bulk lots and repackages them in small quantities, or (2) contact the manufacturer, who can advise you of local sources of supply for the products it manufactures.

Many chemical manufacturers have offices in principal cities. Frequently, they are listed under the product you wish to locate. Also, principal manufacturers are listed, under the product you seek, in the *O.P.D. Chemical Buyer's Directory* (Schuell Publishing Company). Most libraries have a copy.

Every important chemical that is specified in the more than 500 formulas in the fourth edition of *The Formula Manual* is listed in this appendix. After the name of the chemical will be found (1) the common name, if any, (2) the chemical formula, (3) a description of the material, (4) how it is made or what it is derived from, (5) any cautions that should be observed in its use, and (6) a list indicating its source of supply. Following is the index of chemicals.

The following companies handle the ingredients listed in the formulas of this book and are presented as a convenience to our readers. No endorsement is given or implied for any company.

CHEMICALS:Student Science Service, 622 W. Colorado St., Glendale, Calif. 91204, (213)-247-6910.

PERFUMES AND ESSENTIAL OILS: Shemen Tov Corporation, P.O. Box 455 WOB, West Orange, N.J. 07052.

CHEMICAL HOUSES—SALES TO BONAFIDE INSTITUTIONS

Aldrich Chemical Company, Inc.
940 W. St. Paul Ave.
Milwaukee, Wisconsin 53233

Warehouse and shipping facilities in New Jersey.

Sales only to bonafide institutions upon the receipt of a written purchase order. Products are for use by a person trained in chemistry or under the supervision of a responsible chemist. Catalog sent free of charge.

Ventron Corporation

Alfa Products
Home Office and Order Department:
P.O. Box 299
152 Andover Street
Danvers, Massachusetts 01923

Western Order Department and Stock Point:
Ventron Corporation
Alfa Products 2098 Pike Street
San Leandro, California 94577

Canadian Order Department
Ventron Corporation
Alfa Products
1500 Stanley Street
Suite 405 Montreal 110, Quebec

Sales only to bonafide institutions upon the receipt of a written purchase order. Products are for use by a person trained in chemistry or under the supervision of a responsible chemist. Catalog sent free of charge.

CHEMICAL HOUSES SELLING TO INSTITUTIONS AND TO INDIVIDUALS

Tri-Ess Sciences, Incorporated
622 W. Colorado Street
Glendale, California 91204
Telephone: (213) 247-6910

Sales of chemicals, and scientific items and essential oils to institutions and individuals. Small quantities of chemicals and single items of laboratory supplies sold. Customers may visit and purchase from showroom. Orders shipped by United Parcel Service, C.O.D. Minimum order $5.00. Catalogs available for $1.00.

Buckeye Scientific Company
P.O. Box 15181
Columbus, Ohio 43215
Telephone: (614) 276-2273

Sales and chemicals to schools and individuals. Chemicals sold in small amounts. Catalogs available for $2.00. Minimum order $25.00 Shipping and handling prepaid by Buckeye. Visa and Mastercharge accepted. Prices quoted on almost any chemical or type of chemical equipment.

Shemen Tov Corporation
P.O. Box 455 WOB
West Orange
New Jersey 07052

Telephone: (201) 673-2350

Sales of chemicals, essential oils, flavors, fragrances and aromatic chemicals to institutions and individuals. Items supplied in quantities of ounces up to drums. No minimum order charge. No charge for products list. With written inquiries please enclose a long, self-addressed, stamped (30 cents) envelope. Telephone quotations will be given upon request.

Universal Scientific of Arizona, Incorporated
1008 A. East Vista Del Cerro
Tempe, Arizona 85281

(602) 866-2780

Sales of chemicals to institutions and individuals. Chemicals sold in small amounts. No charge for catalog. Minimum order $5.00.

VWR Scientific, Inc.

Albuquerque, New Mexico
(800) 525-9636

Anchorage, Alaska 99501
1301 East First Avenue

Atlanta, Georgia 30324
P.O. Box 13007 Station K

Baltimore, Maryland 21227
6601 Amberton Drive

Denver, Colorado 80239
P.O. Box 39396

Honolulu, Hawaii 96820
P.O. Box 29697

Boston, Massachusetts 02101
P.O. Box 232

Buffalo, New York
P.O. Box 8
Kenmore, New York 14217

Columbus, Ohio 43216
P.O. Box 855

Dallas, Texas 75235
P.O. Box 35106

Portland, Oregon
(503) 225-0440

Rochester, New York 14603
P.O. Box 1050

Houston, Texas 77033
P.O. Box 33348

Kansas City, Missouri 64108
1515 Broadway

Los Angeles
P.O. Box 1004
Norwalk, California 90650

Midland, Missouri 48640
P.O. Box 2210

New York City, New York
(212) 499-6835

Philadelphia, Pennsylvania 19104
3202 Race Street

Phoenix, Arizona 85038
P.O. Box 29027

Pittsburgh, Pennsylvania 15238
147 Delta Drive

Salt Lake City, Utah 84110
P.O. Box 1678

San Diego, California
(714) 262-0711

San Francisco, California 94119
P.O. Box 3200

Seattle, Washington 98124
P.O. Box 3551

Tucson, Arizona
(602) 623-3618

International Department
P.O. Box 3200
San Francisco, California 94119
Cable VANROG

Suppliers of industrial, educational and medical laboratories. Catalog sent free of charge. Orders less than $25.00 may be subject to a handling charge. A large chemical house, but will accept orders from individuals with a minimum $25.00 order.

INDEX OF CHEMICALS

ACACIA (Gum Arabic): White powder or flakes, soluble in water. Dried from the plant Acacia Senegal. Available from retail drugstores, wholesale drug distributors, and industrial chemical suppliers.

ACETIC ACID (Vinegar Acid): Clear colorless liquid, miscible with water, alcohol, glycerin, and ether. Made by oxidation of petroleum gases. Available from retail drugstores, wholesale drug distributors, and industrial chemical suppliers.

ACETONE CH_3COCH_3: Miscible with water, alcohol, ether, chloroform, and most oils. Made by oxidation of cumene. CAUTION: Extremely volatile and flammable. Available from retail drugstores, wholesale drug distributors, industrial chemical suppliers, oil distributors, and repackagers of chemicals.

ACTIVATED CHARCOAL: Black powder. Obtained by the destructive distillation of carbonaceous materials such as wood or nut shells. It is activated by heating to approximately 900° C. with steam or carbon dioxide which produces a honeycomb internal structure, making it highly adsorptive. Available from retail drugstores, wholesale drug distributors, retail and wholesale paint and hardware stores, and industrial chemical suppliers.

ALMOND OIL: White to yellowish oil, distilled from ground kernels of bitter almonds imported from Spain, Portugal, or France. CAUTION: Vapors are toxic. Available from retail drugstores and wholesale drug distributors.

ALUMINUM CHLORIDE, ANHYDROUS AlC_3: White or yellowish crystals. Derived by reaction of purified gaseous chlorine with molten aluminum, by reaction of bauxite with coke and chlorine at about 1600° F. CAUTION: Highly toxic by ingestion and inhalation, strong irritant to tissue. Available from retail drugstores, wholesale drug distributors, and building supply dealers.

ALUMINUM POTASSIUM SULFATE (Alum) $Al2(SO_4)_3.K_2SO_4.24H_2O$: White crystals or powder, soluble in water. Derived from alunite leucite and other minerals. Acts as an astringent. Available from retail drugstores, wholesale drug distributors, and industrial chemical suppliers.

ALUMINUM POWDER Al_2: Gray to silver powder, milled from aluminum or its alloys. Particles are dispersed in a vehicle such as paint. Available from retail and wholesale paint and hardware stores and industrial chemical suppliers.

ALUMINUM SODIUM SULFATE (see alum)

ALUMINUM STEARATE $Al(C_{18}H_{35}02)$: White powder, soluble in petroleum and turpentine oil. Made by reacting aluminum salts with stearic acid. Available from retail drugstores, wholesale drug distributors, and industrial chemical suppliers.

ALUMINUM SULFATE (Alum) $Al_2(SO_4)_3$: White crystals, soluble in water. Made by treating kaolin with sulfuric acid.

AMMONIA, HOUSEHOLD: NH_4OH and H_2O. A dilute solution of ammonium hydroxide. Available from a retail supermarket, wholesale grocery, and feed and grain supplier.

AMMONIUM SELENATE $(NH_4)_2SO_4$: Colorless crystals, soluble in water. CAUTION: May be mildly toxic. Available from retail drugstores, wholesale drug distributors, and industrial chemical suppliers.

AMMONIUM CARBONATE (Hartshorn) $(NH_4)HCO_3$ $(NH_4)CO_2NH_2$: White powder, soluble in cold water. A mixture of ammonium acid carbonate and ammonium carbamate. Derived from the heating of ammonium salts with calcium carbonate. CAUTION: When heated, irritating fumes may result. Available from retail drugstores, wholesale drug distributors, and industrial chemical suppliers.

AMMONIUM CHLORIDE (Sal Ammoniac) NH_4Cl: White crystals, soluble in water and glycerol. Derived from the reaction of ammonium sulfate and sodium chloride solutions. Available from retail drugstores, wholesale drug distributors, and industrial chemical suppliers.

AMMONIUM CITRATE $(NH_4)_2HC_6H_5O_7$: White granules, soluble in water. Available from retail drugstores, wholesale drug distributors, and industrial chemical suppliers.

AMMONIUM HYDROXIDE (Aqua Ammonium) NH_4OH: Water solution of ammonia gas. CAUTION: Toxic by ingestion. Liquid and vapors may be irritating to eyes and skin. Available from industrial chemical suppliers.

AMMONIUM OLEATE (Ammonia Soap) $C_{17}H_{33}COONH_4$: Brown, jellylike mass, soluble in water and alcohol. Acts as an emulsifying agent. Available from retail drugstores, wholesale drug distributors, and industrial chemical suppliers.

AMMONIUM NITRATE (Saltpeter) NH_4NO_3: Colorless crystals, soluble in water, alcohol, and alkalies. Made by the action of ammonia vapor on nitric acid. CAUTION: Do not store in high temperatures. Available from farm suppliers, industrial chemical suppliers and repackagers of chemicals.

AMMONIUM PHOSPHATE $(NH_4)_2HPO_4$: White crystals moderately soluble in water. Derived from the interaction of phosphoric acid and ammonia. Available from wholesale drug distributors, feed and grain supply, and industrial chemical suppliers.

AMMONIUM STEARATE $C_{17}H_{35}COONH_4$: Tan, waxlike solid, dispersible in hot water, soluble in hot toluene. Available from retail drugstores, wholesale drug distributors, industrial chemical suppliers, and repackagers of chemicals.

AMMONIUM SULFATE $(NH_4)_2SO_4$: Gray to white crystals, soluble in water. Made by neutralizing synthetic ammonia with sulfuric acid. Available from retail drugstores, wholesale drug distributors, industrial chemical suppliers, and repackagers of chemicals.

AMYL ACETATE (Banana Oil) $CH_3COOC_5H_{11}$: Volatile bananalike smell. CAUTION: Toxic. Available from drug store and chemical suppliers.

AMYL MERCAPTAN CH_2: $CHCH_2SH$ (2 propene-thiol): Soluble in alcohol and ether, usually with strong disagreeable odor. Available from chemical repackagers and industrial chemical suppliers.

ANHYDROUS LANOLIN (Wool Fat): Brown jelly, miscible with water. Soluble in benzene, ether, acetone, and slightly soluble in cold alcohol. Available from retail drugstores, wholesale drug distributors, and industrial chemical suppliers.

ANITERACENE OIL: A medium viscosity oil obtained as a coal tar fraction. Available from oil distributors, and repackagers of chemicals.

ANTHRACENE OIL: A coal tar fraction. CAUTION: Hazardous, toxic, and an irritant. Available from oil distributors and repackagers of chemicals.

ANTIMONY CHLORIDE SbOCL: White powder soluble in hydrochloric acid and alkali tartrate solutions. CAUTION: Highly toxic. Available from chemical suppliers.

ANTIMONY POTASSIUM TARTRATE (Tartar Emeric) $K(SbO)C_4H_4O_6$. $1/2H_2O$: White powder, soluble in water. Derived by heating antimony trioxide with a solution of potassium bitartrate and then crystalized. CAUTION: Toxic if taken internally. Available from chemical suppliers.

ASBESTOS POWDER: Gray fibrous powder. Mined as a natural mineral. CAUTION: Do not inhale dust. Available from retail and wholesale paint and hardware stores.

ASPHALT (Residual Oil, Petroleum Asphalt, Trinidad Pitch, Mineral Pitch): Solid to semisolid lumps, turns to viscous liquid at 200° F. Available from retail and wholesale paint and hardware stores and building supply dealers.

BALL CLAY: Tan-colored, powdered, highly plastic with strong bonding power. Mined in various sections of the United States. Available from ceramic shops and repackagers of chemicals.

BARIUM SULFIDE (Black Ash) BaS: Yellowish green to gray powder, soluble in water. Made by roasting barium sulfate and coal together, adding water, and evaporating. CAUTION: Toxic if taken internally. Available from industrial chemical suppliers and wholesale hardware dealers.

BAY RUM: Amber liquid made from a mixture of bay oil, orange peel oil, oil of pimenta, and alcohol. Available from retail drugstores, wholesale drug distributors, industrial chemical suppliers, and repackagers of chemicals.

BEEF TALLOW: Solid fatty material found in beef. Available from retail supermarkets.

BEESWAX: See White Beeswax.

BENTONITE (Sodium Bentonite): Light powder, insoluble in water, expands to many times its size in water. Mined in Wyoming, Mississippi, Texas, Canada, Italy, and Russia. Available from industrial chemical suppliers.

BENZENE C_6H_6: Colorless liquid, made by the catalytic reforming of petroleum, and also by the fractional distillation of coal tar. CAUTION: Flammable. Available from retail and wholesale paint and hardware stores, oil distributors, industrial chemical suppliers, and repackagers of chemicals.

BENZENE HEXACHLORIDE (BHC, Lindane): Insecticide. CAUTION: Highly toxic.

BORAX (Sodium Borate) $NA_2B_4O_7.IOH_2O$: White powder, soluble in water. Mined in the western United States. Available from retail and wholesale paint and hardware stores, retail and wholesale groceries, industrial chemical suppliers, and repackagers of chemicals.

BORIC ACID (Boracic Acid) H_3BO_3: Colorless, odorless, white powder, soluble in water, alcohol, and glycerine. Made by the addition of hydrochloric or sulfuric acid to a borax solution, and then crystallizing. Available from retail drugstores, wholesale drug distributors, industrial chemical suppliers, and repackagers of chemicals. CAUTION: Toxic by ingestion in undiluted form.

BENZOIC ACID (Carboxybenzene C_6H_5COOH: White scales or needlelike crystals, soluble in alcohol, ether, chloroform, benzene, and turpentine. Made by the oxidation of toluene. Available from retail drugstores, wholesale drug distributors, industrial chemical suppliers, and repackagers of chemicals.

BERGAMOT OIL: Honey-colored oil, soluble in alcohol. Derived from the fruits of *Citrus Bergamia Risso et Painteau*. Available from retail drugstores, wholesale drug distributors, and repackagers of chemicals.

BURGUNDY PITCH: A tacky liquid, soluble in alcohol and acetone. Extracted from Norway spruce. Available from retail drugstores, wholesale drug distributors, industrial chemical suppliers, and repackagers of chemicals.

BUTYL ALCOHOL (Butanol) $CH_3(CH_2)_2CH_2OH$: Colorless liquid, soluble in water, miscible with alcohol and ether. Made by the hydrogenation of butylraldehyde, obtained in the Oxo process.

CADMIUM ACETATE $Cd(OOCH_3)_2.3H_2O$: Colorless crystals, soluble in water. CAUTION: Toxic. Available from retail drug stores, wholesale drug distributors, industrial chemical suppliers, and repackagers of chemicals.

CALCINED MAGNESIA (Magnesite) MGO: Derived by firing magnesite up to 1450° C., at which time it converts to an adsorptive medium with high internal porosity. Available from retail drug stores, wholesale drug distributors, industrial chemical suppliers, and repackagers of chemicals.

CALCIUM CARBONATE (Chalk) $CACO_3$: White powder slightly soluble in water, highly soluble in acids. Derived principally from limestone. Available from retail drug stores, wholesale drug distributors, industrial chemical suppliers, repackagers of chemicals, feed and grain suppliers, and building supply dealers.

CALCIUM CHLORIDE $CaCl_2$: White flakes that decompose in water. Absorptive agent. Available from retail drugstores, wholesale drug distributors, industrial chemical suppliers, and repackagers of chemicals.

CALCIUM HYPOCHLORITE (Chlorinated Lime) $Ca(OCL)_2$: White crystalline solid, soluble in water. Derived from the chlorination of a lime/caustic slurry. CAUTION: Toxic. Available from lumber distributors, and industrial chemical suppliers.

CALCIUM LACTATE $Ca(C_3H_5O_3)_2.5(H_2O)$: White, tasteless powder. Soluble in water. Available from retail drugstores, wholesale drug distributors, industrial chemical suppliers, oil distributors, and repackagers of chemicals.

CALCIUM SULFATE $CaSO_4$: White, odorless crystals or powder. Only slightly soluble in water. Occurs in nature as a hydrated form of gympsum, and also as an anhydrate. Available from building supply dealers and repackagers of chemicals.

CALCIUM SULFATE (Plaster of Paris) $CaSO_4.H_2O$: White solid, formed by reaction of soluble calcium salt solution and sodium sulfate solution. Available through chemical supply houses and chemical repackagers.

CAMPHOR (Gum Camphor, Camphanone) $C_{10}H_{16}O$: Colorless or white crystals, soluble in alcohol. Derivation: steam distillation of camphor tree wood. CAUTION: Vapors flammable. Available from retail drugstores, wholesale drug distributors, industrial chemical suppliers, and repackagers of chemicals.

CAMPHOR OIL: Pale yellow oily liquid. Made by distilling the flowers of the Canaga odorata that grows in Java. Available from retail drugstores, wholesale drug distributors, industrial chemical suppliers, and repackagers of chemicals.

CARBON DISULFIDE CS_2: Clear liquid, soluble in alcohol, benzene, and ether. CAUTION: Highly flammable. Available from retail drugstores, wholesale drug distributors, industrial chemical suppliers, and repackagers of chemicals.

CARBORUNDUM: Trade name for a full line of abrasives made by the Carborundum Company. Available from building supply dealers.

CARNAUBA WAX (Brazil Wax): Yellow to brown hard lumps, melting point 84°-86° C. Collected from the leaves of the Brazilian wax palm, *Copernica cerifera*. Available from oil distributors, industrial chemical suppliers, and repackagers of chemicals.

CARBOLIC ACID (Phenol) C_6H_5OH: Soluble in water and alcohol. Made by the oxidation of cumene. CAUTION: Toxic by ingestion, inhalation, and skin absorption. Available from retail drug stores, wholesale drug distributors, industrial chemical suppliers, and repackagers of chemicals.

CARBON BLACK (Stove Black, Furnace Black): Black insoluble amorphous powder. Made by the incomplete combustion of natural gas or petroleum. Available from retail paint and hardware stores, industrial chemical suppliers, ceramic shops, and repackagers of chemicals.

CASEIN ADHESIVE: An adhesive made from casein which is a protein percipitated from milk. Available from wholesale drug suppliers, chemical supply houses, chemical repackagers, retail and wholesale paint and hardware supply stores.

CASTILE SOAP: Olive oil is used for Castile soap; transparent soaps are made from decolorized fats. Available from retail drugstores, wholesale drug distributors, and retail and wholesale grocery stores.

CASTILE SOAP CHIPS: A soap made from olive oil and soda that has been cut into small chips. Available through retail and wholesale drug distributors retail supermarkets and wholesale grocery stores.

CASTOR OIL: A non-drying oil used in medicine and also used extensively in industry; from the plant *Ricinus communis*. Available from retail and wholesale drug and hardware stores, oil companies, retail paint stores and chemical repackagers.

CASTOR OIL (Ricinus Oil): Pale yellow oil, soluble in alcohol. Derived from pressing the seeds of the castor bean, *Ricinus communis*. Available from retail drugstores, wholesale drug distributors, industrial chemical suppliers, and repackagers of chemicals.

CAUSTIC POTASH (Potassium Hydroxide) KOH: White flakes, soluble in alcohol, water, or glycerin. Made by electrolysis of a potassium chloride solution. CAUTION: Heats on contact with water, can cause severe burns to skin. Handle with care. Store in airtight container. Available from retail paint and hardware stores, industrial chemical suppliers, ceramic shops, and repackagers of chemicals.

CAUSTIC SODA (Sodium Hydroxide) NAOH: White chips, soluble in water or alcohol. Made by electrolysis of a sodium chloride solution. CAUTION: Heats on contact with water, can cause severe burns to skin. Handle with care. Store in airtight container. Available from retail drugstores, wholesale drug distributors, retail and wholesale paint and hardware stores, industrial chemical suppliers, and repackagers of chemicals.

CEDAR OIL: An aromatic essential oil extracted from cedar bark and wood. Available from chemical supply houses and oil of essence importers.

CEDAR OIL EMULSION: Cedar oil emulsified with water, using any suitable emulsifier such as liquid detergent. Available from chemical supply houses and oil of essence importers.

CELLULOSE STEARATE: A stearate used in the manufacture of soaps and industrial chemicals obtained from plant cellulose of various origins. Available from repackagers of chemicals, and industrial chemical suppliers.

CERESIN WAX (Ozocerite, Mineral Wax): White or yellow solid, soluble in alcohol. Melting point 68°-72° C. Made by purifying ozocerite with sulfuric acid and then filtering through charcoal. Available from industrial chemical suppliers, ceramic shops, and repackagers of chemicals.

CETYL ALCOHOL $C_{16}H_{33}OH$: White crystals, soluble in alcohol. Melting point 49° C. Made by saponifying spermaceti with caustic alkali. Available from retail drugstores, wholesale drug distributors, industrial chemical suppliers, and repackagers of chemicals.

CETYLTRIMETHYLAMMONIUM BROMIDE $C_{16}H_{33}(CH_3)_3NBr$ (Ammonium Salt): White powder, soluble in water or alcohol. Has surface-active and germicidal properties. Available from retail drugstores, wholesale drug distributors, industrial chemical suppliers, and repackagers of chemicals.

CHLORAL HYDRATE $CClz.CHO.H_2O$: Occurs in crystalline form, soluble in water; formed by passing chlorine through absolute alcohol, dangerous if taken internally in large doses causing failure of the circulation or respiration system. Available from wholesale drug suppliers, chemical repackagers and industrial chemical suppliers.

CHLORIMINE: An industrial organic compound, soluble in ether, alcohol and acetone; used as an oxidizing agent and a germicide. Available through chemical supply houses and chemical repackagers.

CHLORINATED LIME (Bleaching Powder) $CaCL(CIO).4H_2O$: White granules that decompose in water. Made by reacting chlorine with slaked lime. CAUTION: Forms chlorine when mixed with water. Available from industrial chemical suppliers, building supply dealers, and repackagers of chemicals.

CHLOROPHYLL $C_{55}H_{70}MgN_4O_6$: Green material found in plants and algae. Can be had in aqueous, alcoholic, or oil solutions. Made by extraction from the plant source. Available from retail drugstores, wholesale drug distributors, industrial chemical suppliers, and repackagers of chemicals.

CHROMIC OXIDE (Chromium Oxide, Chromia, Chromium Sesquioxide, Green Cinnabar) Cr_2O_3: Bright-green crystalline powder, insoluble in water, acids, and alkalies. Derived by heating chromium hydroxide; dry ammonium dichromate, and sodium dichromate with sulfur and washing out the sodium sulfate. Available from industrial chemical suppliers, building supply dealers, and repackagers of chemicals.

CINNAMON OIL (Cassia Oil): Light yellow aromatic oil, soluble in alcohol. Distilled from the leaves and twigs of the plant *Cinnamonum cassia*. Available from retail drugstores, wholesale drug distributors, and repackagers of chemicals.

CITRIC ACID $HOOCCH_2C(OH)(COOH)CH_2COOH.H_2O$: White crystals, soluble in water or alcohol. Derived by mold fermentation from lemon, lime, pineapple juice, and molasses. Available from retail drugstores, wholesale drug distributors, industrial chemical suppliers, and repackagers of chemicals.

CITRONELLA OIL: Light yellow essential oil, soluble in alcohol. Derived by the steam distillation of the grass *Cymbopogon Nardus*. CAUTION: Mildly toxic if taken internally. Available from retail drugstores, wholesale drug distributors, industrial chemical suppliers, and repackagers of chemicals.

CLAY (Hydrated Aluminum Silicate) $Al_2O_3SIO_2H_2O$: Tan powder, ranging in particle size from 150 to less than 1 micron. Absorbs water to form a plastic mass. Derived from nature by natural weathering, crushing, and screening of rock. Available from industrial chemical suppliers, building supply dealers, ceramic shops, and repackagers of chemicals.

CLOVES: The dried flowers of *Eugenia aromatica*. Flowers are frequently ground to a powder, or distilled to produce oil of clove. Available from retail drugstores, wholesale drug distributors, retail supermarkets, wholesale groceries, and repackagers of chemicals.

COAL TAR: Black viscous liquid (or semisolid), napthalenelike odor; sharp burning taste; obtained by destructive distillation of coal. Soluble in ether, benzene, carbon disulfide, chloroform; partially soluble in alcohol, acetone, methanol, benzene; slightly soluble in water. CAUTION: Highly toxic by inhalation.

COBALT CHLORIDE (Cobaltus Chloride) $CoCl_2.6H_2O$: Blue or red crystals. Soluble in water, alcohol, or acetone. Made by the action of hydrochloric acid on cobalt. Available from retail drugstores, wholesale drug distributors, industrial chemical suppliers, and repackagers of chemicals.

COCONUT OIL: White, semisolid, lardlike fat. Soluble in alcohol. Melting point 83° F. Made by press extraction of coconut meat followed by alkali refining. Available from retail drugstores, wholesale drug distributors, industrial chemical suppliers, and repackagers of chemicals.

COCOA BUTTER (Theobroma Oil): Yellow-white solid. Melting point 30°-35° C. Soluble in ether or chloroform. Made by the expression of cocoa beans and solvent extraction. Available from retail drugstores, wholesale drug distributors, industrial chemical suppliers, and repackagers of chemicals.

COPPER CHLORIDE Cu Cl: White solid, insoluble in water, formed by reaction of cupric chloride solution and copper metal. Available from repackagers of chemicals and industrial chemical suppliers.

COPPER NAPHTHENATE: A green-blue solid, soluble in gasoline, benzene, and mineral oil distillates. Made by combining cupric to a solution of sodium naphthenate. CAUTION: Mildly toxic by ingestion. Available from retail drugstores, wholesale drug distributors, industrial chemical suppliers, and repackagers of chemicals.

COPPER SULFATE (Blue Vitriol, Bluestone) $CuSO_4.5H_2O$: Blue crystals, lumps, or powder. Soluble in water or methanol. Made by the action of dilute sulfuric acid on copper or its oxides. CAUTION: Highly toxic. Available from retail drugstores, wholesale drug distributors, industrial chemical suppliers, and repackagers of chemicals.

CORN OIL (Maize Oil): Pale yellow liquid, partially soluble in alcohol. The germ is removed from the kernel and cold pressed. Available from retail supermarkets, wholesale groceries, industrial chemical suppliers, and repackagers of chemicals.

CORN SYRUP (Glucose): Viscous liquid consisting of a mixture of dextrose, maltose, and dextrius with about 20% water. Soluble in water and glycerine. Made by the hydrolysis of starch and the action of hydrochloric acid. Available from retail supermarkets, wholesale groceries, industrial chemical suppliers, and repackagers of chemicals.

COTTONSEED OIL: Pale yellow to clear oil, soluble in ether, benzene, chloroform, and carbon disulfide. Made by solvent extraction or hot pressing of cotton seeds. Available from retail drugstores, wholesale drug distributors, industrial chemical suppliers, and repackagers of chemicals.

CRESOL (Methyl Phenol) $CH_3C_6H_4OH$: Colorless or yellowish to pinkish liquid. Soluble in alcohol, glycol, and dilute alkalies. CAUTION: Toxic and irritant. Available from retail drugstores, wholesale drug distributors, industrial chemical suppliers, and repackagers of chemicals.

CREOSOTE (Wood Tar, Beechwood): Colorless oil liquid, miscible with alcohol or ether. A mixture of phenols obtained by the destructive distillation of wood tar. Available from industrial chemical suppliers, building supply dealers, and repackagers of chemicals.

CRESYLIC ACID: A commercial mixture of phenolic materials, made from petroleum or coal tar. CAUTION: Toxic, absorbed through skin. Use gloves to handle. Available from retail drugstores, wholesale drug distributors, industrial chemical suppliers, and repackagers of chemicals.

DENATURED ALCOHOL: Ethyl alcohol that has been contaminated with a minute amount of another material to make it unfit for human consumption as a beverage. Clear white liquid. CAUTION: May be toxic if taken internally. Flammable. Available from retail drugstores, wholesale drug distributors, industrial chemical suppliers, and repackagers of chemicals.

DEODORIZED KEROSENE: Kerosene that has been treated chemically to mask its odor. CAUTION: Toxic if taken internally. Flammable. Available from service stations, petroleum suppliers and industrial chemical suppliers.

DERRIS POWDER: See Pyrethrum.

DERRIS ROOT: The root of the shrub *Malacceusis.* Chief active ingredient is rotenone. Available from industrial chemical companies.

DIBASIC AMMONIUM PHOSPHATE (Ammonium Phosphate, Secondary; Diammonium Hydrogen Phosphate; Diammonium Phosphate; DAP) $(NH_4)_2 HPO_4$: White crystals or powder, milky alkaline in reaction, soluble in water. Available from retail drugstores, wholesale drug distributors, industrial chemical suppliers, and repackagers of.chemicals.

DIATOMACEOUS EARTH (Keiselguhr, Diatomite): A bulky light material containing 88% silica. The balance is made up of the skeletons of small prehistoric plants related to algae. Can be had in either brick or powder form. Available from industrial chemical suppliers, building supply dealers, ceramic shops, and repackagers of chemicals.

DIGLYCOL LAURATE (Diethylene Glycol Monolaurate) $C_{11}H_{23}COOC_2H_4OC_2H_4 OC_2H_4OH$: Straw-colored, oily liquid, nontoxic. Dispersible in water. Derived from thelauric acid ester of diethylene glycol. Available from retail drugstores, wholesale drug distributors, industrial chemical suppliers, and repackagers of chemicals.

DIGLYCOL OLEATE (Diethylene Glycol Monooleate) $C_{17}H_{33}COOC_2H_4OC_2H_4 OH$: Light red, oily liquid; fatty odor. Soluble in ethanol, naptha, ethyl acetate, methanol; partly soluble in cottonseed oil; insoluble in water. Combustible; nontoxic. Available from retail drugstores, wholesale drug distributors, industrial chemical suppliers and repackagers of chemicals.

DIGLYCOL STEARATE $(C_{17}H_{35}COOC_2H_4)20$: White, waxlike solid. Disperses in hot water, soluble in hot alcohol. Made by using stearic acid and the ester of diethylene glycol. Available from retail drugstores, wholesale drug distributors, industrial chemical suppliers and repackagers of chemicals.

DIMETHYLMORPHOLINE $OCH(CH_3)CH_2NHCH_2CH(CH_3)$: Liquid. Flash point 112° F. CAUTION: Flammable. Available from retail drugstores, wholesale drug distributors, industrial chemical suppliers and repackagers of chemicals.

ESSENTIAL OILS: Volatile oils derived from the leaves, stems, flowers, and twigs of plants, and from the rinds of fruits. Methods of extraction are by steam distillation, pressing fruit rinds, solvent extraction, and maceration of flowers and leaves. Generally soluble in alcohol and chloroform. Available from importers of essential oils.

ETHYL ACETATE (Acetic Ether; Acetic Ester, Vinegar Naphtha) $CH_3COOC_2H_5$: Colorless, fragrant liquid. Soluble in chloroform, alcohol, and ether; slightly soluble in water. CAUTION: Moderately toxic by inhalation and skin absorption. Irritating to eyes and skin. Highly combustible. Available from retail drugstores, wholesale drug distributors, retail and wholesale paint and hardware stores, industrial chemical suppliers, and repackagers of chemicals.

ETHYL ALCOHOL (VODKA): See Appendix 4.

ETHYLENE CHLORIDE (Ethylene Dichloride) $ClCH_2CH_2Cl$: Colorless, oily liquid, miscible with most organic solvents. Made by the action of chlorine or ethylene. CAUTION: Toxic by ingestion, inhalation, and skin absorption. Irritant to eyes and skin. Handle with care. Available from wholesale drug distributors, industrial chemical suppliers, and repackagers of chemicals.

ETHYLENE DICHLORIDE $C_2H_4Cl_2$: Formed by reacting ethylene with chlorine. Available from repackagers of chemicals and industrial chemical suppliers.

ETHYLENE GLYCOL (Glycol) CH_2OHCH_2OH: Clear, colorless, syrupy liquid, soluble in water and alcohol. Made from formaldehyde, water, and carbon monoxide with hydrogenation of the resulting glycolic acid. Available from retail drugstores, wholesale drug distributors, industrial chemical suppliers, and repackagers of chemicals.

ETHYLENE GLYCOL MONOETHYLETHER (Cellosolve) $HOCH_2OC_2H_5$: Colorless liquid, miscible with water and hydrocarbons. Flash point 120° F. CAUTION: Flammable. Available from chemical supply houses, chemical repackagers and wholesale hardwares.

EUCALYPTUS OIL: A distillation product of the leaves from the eucalyptus tree, *Eucalyptus globulus;* used in the manufacture of antiseptics and cosmetics. Available from wholesale drug distributors, chemical repackagers and chemical supply houses.

FERRIC CHLORIDE (Iron Chloride) $FeCl_3$: A black-brown solid in water, alcohol, and glycerol. Made by the action of chlorine on ferrous sulfate. Available from retail drugstores, wholesale drug distributors, industrial chemical suppliers, and repackagers of chemicals.

FERRIC OXIDE: Iron mineral appearing natural. Available from chemical supply houses.

FISH OIL: A drying oil obtained from salt-water fish. Used in soap making. Available from paint suppliers, industrial chemical suppliers, and hardware suppliers.

FLOWER OF SULFUR: A percipitated form of purified sulfur produced by various means of commercial processes. Available from retail paint and hardware stores, ceramic shops, retail supermarkets, drug repackagers and drugstores.

FLOWERS OF SULFUR: Yellow crystals or powder, partially soluble in alcohol. Melting point 112° C. Mined in various sections of the United States. Available from retail drugstores, wholesale drug distributors, industrial chemical suppliers, and repackagers of chemicals.

FORMALDEHYDE HCHO: An aqueous solution. Made by oxidation of synthetic methanol. CAUTION: Highly toxic by ingestion, inhalation, or skin contact. Available from retail drugstores, wholesale drug distributors, industrial chemical suppliers, and repackagers of chemicals.

FUEL OIL (Furnace Oil): Number 1 or 2 grade. Oil used in home-heating furnaces. Available from service stations and oil distributors.

FULLER'S EARTH: A porous colloidal aluminum silicate of 1 micron or less, having high adsorptive power. Mined in Florida, England, and Canada. Available from industrial chemical suppliers, building supply dealers, ceramic shops, and repackagers of chemicals.

FURFURALDEHYDE (Bran Oil) C_4OCHO: Colorless liquid, soluble in water. Derived from grain hulls. CAUTION: Highly toxic, can be absorbed through skin. Available from retail drugstores, wholesale drug distributors, oil distributors, industrial chemical suppliers, and repackagers of chemicals.

GELATIN: White to yellow powder, soluble in hot water. Made by boiling animal by-products with water. Will absorb up to 10 times its weight of water. Available from retail and wholesale groceries and repackagers of chemicals.

GLUCOSE: Starch or corn syrup. Viscous, syrupy liquid. A mixture of dextrose, maltose, and dextrin. Colorless to yellowish, soluble in water and alcohol. Available from grocery stores, chemical repackagers and industrial chemical suppliers.

GLYCERIN (Glycerol) $C_3H_3(OH)_3$: A clear, colorless, syrupy liquid, soluble in water and alcohol. Made by the hydrogenation of carbohydrates with a nickel catalyst. Available from retail drugstores, wholesale drug distributors, and industrial chemical suppliers.

GUAIACOL $OHC_6H_4OCH_3$: Yellowish, oily liquid or crystals. Soluble in alcohol, moderately soluble in water. Made by extracting beechwood creosote with alcoholic potash. CAUTION: Moderately toxic. Available from chemical supply houses.

HEXACHLORODIPHENYL OXIDE $C_{12}H_4Cl_6O$: Light yellow liquid, soluble in methanol ether. Available from retail drugstores, wholesale drug distributors, industrial chemical suppliers, and repackagers of chemicals.

HEXAMETHYLENETETRAMINE $(CH_2)_6N_4$: White crystalline powder, soluble in water. Made by the action of ammonia on formaldehyde. CAUTION: Moderately toxic, flammable. Available from retail drugstores, wholesale drug distributors, industrial chemical suppliers, and repackagers of chemicals.

HOUSEHOLD BLEACH: Aqueous solution of chlorine made by passing chlorine gas through water or adding solid chlorine to water. Available from retail supermarkets.

HYDRATED LIME (Calcium Hydroxide) $Ca(OH)_2$: White powder, soluble in glycerin. Made by the action of water on calcium oxide. CAUTION: Skin irritant. Available from retail drugstores, feed and grain suppliers, oil distributors, industrial chemical suppliers, and repackagers of chemicals.

HYDROCHLORIC ACID (Muriatic Acid): Hydrogen chloride in water solution. Derived as a by-product of the chlorination of benzene. CAUTION: Highly toxic by ingestion and inhalation, can be absorbed by skin. Strong irritant to eyes and skin. Available from pool supply companies, hardware suppliers and chemical suppliers.

HYDROFLUORIC ACID (Hydrogen Fluoride in Aqueous Solution): Colorless, fuming mobile liquid. Will attack glass and any silica-containing material. CAUTION: Highly corrosive to skin and mucous membranes; highly toxic by ingestion and inhalation. Use with extreme caution. Wear rubber gloves and use in well-ventilated area. Available from retail drugstores, wholesale drug distributors, industrial chemical suppliers, and repackagers of chemicals.

HYDROGEN PEROXIDE: Colorless dilute aqua solution. CAUTION: Highly toxic in concentrated form. Relatively low toxicity in dilute aqua solution. Sold as 3% solution over drugstore counter as a general antiseptic. Available from retail drugstores and wholesale drug distributors.

INDIA INK: Carbon black with shellac, borax, or soap in water. Available from retail drugstores, wholesale drug distributors, retail and wholesale paint and hardware stores, and retail and wholesale groceries.

IRON CARBONYL (Iron Pentacarbonyl) $Fe(CO)_5$: Yellow liquid, soluble in organic solvents. Made by treating iron dust with carbon monoxide in the presence of the catalyst ammonia. Available from retail drugstores, wholesale drug distributors, industrial chemical suppliers, and repackagers of chemicals.

IRON OXIDE (Jeweler's Rouge) Fe_2O_3: Reddish-brown fine powder, soluble in acids. Made by the interaction of a solution of ferrous sulfate and sodium carbonate. Available from industrial chemical suppliers, ceramic shops, and repackagers of chemicals.

IRON SULFATE (Ferric Sulfate) $Fe_2(SO_4)_3 \cdot 9H_2O$: Yellow crystals, soluble in water. Made by the addition of sulfuric acid to ferric hydroxide. Available from industrial chemical suppliers, ceramic shops, and repackagers for chemicals.

ISOPROPYL ALCOHOL (Isopropanol) $(CH_3)_2CHOH$: White, sweet-smelling liquid. Soluble in water, ether, or alcohol. Made by treating propylene with sulfuric acid and then hydrolyzing. CAUTION: Mildly toxic by inhalation and ingestion. Flammable. Available from retail drugstores, wholesale drug distributors, and industrial chemical suppliers.

JAPAN WAX (Sumac Wax): Pale yellow solid, soluble in benzene and naphtha. Melting point 53° C. Available from industrial chemical suppliers and repackagers of chemicals.

JEWELER'S ROUGE: See Iron Oxide.

JUNIPER OIL (Juniper Tar, Juniper Tar Oil): Thick, clear dark-brown liquid, tarry odor; burning, bitter taste. Soluble in ether, glacial acetic acid, chloroform; partly soluble in alcohol; very slightly soluble in water. Derived by distillation of the wood of *Juniperus oxycedrus*. Available from retail drugstores, wholesale drug distributors, industrial chemical suppliers, and repackagers of chemicals.

KEROSENE: Oily liquid distilled from petroleum. CAUTION: Toxic if taken internally. Flammable. Available from retail and wholesale paint and hardware stores and oil distributors.

LACTIC ACID (Milk Acid) $CH_3CHOHOOH$: Colorless viscous liquid, miscible with water, alcohol, and glycerin. Made by hydrolysis of lactronitrile. Available from wholesale drug distributors, retail paint and hardware stores, and wholesale groceries.

LAMPBLACK: A black or gray pigment made by burning low-grade heavy oils or similar carbonaceous materials with insufficient air, and in a closed system such that soot can be collected in settling chambers. Strongly hydrophobic. Available from oil distributors, industrial chemical suppliers, and building supply dealers.

LANOLIN (Wool Fat): Yellow to light-gray semisolid, soluble in ether or chloroform. Extracted from raw wool and refined. Available from retail drugstores, wholesale drug distributors, industrial chemical suppliers, and repackagers of chemicals.

LARD: Purified internal fat of the hog. Soft white unctuous mass, faint odor, bland taste. Soluble in ether, chloroform, light petroleum hydrocarbons, carbon disulfide; insoluble in water. Available from retail and wholesale groceries.

LATEX (Liquid Rubber): A white, free-flowing liquid obtained from certain species of trees and shrubs. Usually emulsified with water. Available from industrial chemical suppliers, repackagers of chemicals, and hobby shops.

LAURYL PYRIDINIUM CHLORIDE $C_5H_5NClC_{12}H_{25}$: Mottled tan semisolid, soluble in water. CAUTION: May be mildly irritating to skin. Available from retail drugstores, wholesale drug distributors, industrial chemical suppliers, and repackagers of chemicals.

LAVENDER OIL: Essential oil; colorless, yellowish or greenish-yellow; sweet odor; slightly bitter taste. Steam distilled from fresh flowering tops of *Lavandula officinalis*. Available from retail drugstores, wholesale drug distributors, industrial chemical suppliers, and repackagers of chemicals.

LEAD Pb: Heavy soft-gray metal solid, soluble in dilute nitric acid. Made by the roasting of lead sulfide, lead sulfate, and carbonate. CAUTION: Poison. Available from industrial chemical suppliers and repackagers of chemicals.

LECITHIN $CH_2(R)CH(R)CH_2OPO(OH)O(CH_2)_2N(OH)(CH_3)^3$: Light brown to brown semiliquid, partly soluble in water, soluble in chloroform and benzene. Made from soybean oil, corn oil, egg yolk, and vegetable seeds.

LEMON OIL: Yellow liquid; soluble in alcohol, vegetable oils, and mineral oil. Expressed from the peel of lemons. Available from retail drugstores, wholesale drug distributors, industrial chemical suppliers, and repackagers of chemicals.

LINSEED OIL: Amber to brown oil, soluble in alcohol. Made by refining raw linseed oil. Warning: Dries when exposed to air. Keep in airtight container. Available from retail and wholesale paint and hardware stores, industrial chemical suppliers, and repackagers of chemicals.

LUBRICATING OIL: Amber to red liquid of varying viscosity, refined from crude petroleum oil. Available from service stations and oil distributors.

MAGNESIUM CARBONATE $MgCO_3$: White, bulky, light powder; soluble in acids, insoluble in water and alcohol. Made by a union of magnesium sulfate and sodium carbonate solutions. Available from retail drugstores, wholesale drug distributors, industrial chemical suppliers, and repackagers of chemicals.

MAGNESIUM CHLORIDE: White, deliquescent crystals, soluble in water. Available from retail drugstores, wholesale drug distributors, industrial chemical suppliers, and repackagers of chemicals.

MAGNESIUM FLUOSILICATE (Magnesium Silicofluoride) $MgSiF_6.6H_2O$: White efforescent crystalline powder, soluble in water. Derived by treating magnesium hydroxide or carbonate with hydrofluosilicic acid. CAUTION: Highly toxic, strong irritant. Handle with rubber gloves in well-ventilated area only. Available from retail drugstores, wholesale drug distributors, industrial chemical suppliers, and repackagers of chemicals.

MAGNESIUM OXIDE MgO: White solid, formed by heating magnesium carbonate to high temperature; absorbs carbon dioxide from the air to form magnesium carbonate. Available from industrial chemical suppliers and chemical repackagers.

MAGNESIUM STEARATE $Mg(C_{18}H_{35}O_2)2$: Light, white powder, insoluble in water or alcohol. Available from retail drugstores, wholesale drug distributors, industrial chemical suppliers, oil distributors, and repackagers of chemicals.

MAGNESIUM SULFATE (Epsom Salts) $MgSO_4$: Colorless crystals, soluble in water and glycerol. Made by the action of sulfuric acid on magnesium oxide. Available from retail drugstores, wholesale drug distributors, industrial chemical suppliers, and repackagers of chemicals.

MANGANESE CHLORIDE $MnCl_2$: Rose-colored crystals, soluble in water. Slightly toxic. Available from industrial chemical suppliers and chemical repackagers.

METHANOL: See Appendix 4.

MENTHOL (Peppermint Camphor) $CH_3C_6H_9(C_3H_7)OH$: White crystals with strong mint odor; soluble in alcohol, petroleum solvents, and glacial acetic acid. Crystals are formed as a result of freezing mint oil. Available from retail drugstores, wholesale drug distributors, industrial chemical suppliers, and repackagers of chemicals.

MERCURIC CHLORIDE $HgCl_2$: White crystals or powder; soluble in water, alcohol, or ether. CAUTION: Highly toxic by ingestion, inhalation, and skin absorption. Available from retail drugstores, wholesale drug distributors, oil distributors, industrial chemical suppliers, and repackagers of chemicals.

METHYLCELLULOSE (Cellulose Methyl Etiehr; "Methocel"): Grayish white, fibrous powder; aqueous suspensions neutral to litmus. Swells in water to a colloidal solution. Insoluble in alcohol, ether, chloroform, and in water warmer than 123° F. Soluble in glacial acetic acid, unaffected by oils and greases. Derived from cellulose by conversion to alkali cellulose and then, reacting this, with methyl chloride, dimethyl sulfate, or methyl alcohol and dehydrating agents. Available from retail drugstores, wholesale drug distributors, industrial chemical suppliers, and repackagers of chemicals.

METHYL SALICYLATE (Wintergreen Oil) $C_6H_4(OH)COOCH_3$: Colorless or yellow or reddish liquid, soluble in alcohol and glacial acetic acid. Made by heating methanol and salicylic acid in the presence of sulfuric acid. CAUTION: Highly toxic in concentrated form. Available from retail drugstores, wholesale drug distributors, industrial chemical suppliers, and repackagers of chemicals.

MINERAL OIL, WHITE (Liquid Petrolatum): Colorless transparent oil, distilled from petroleum. Available from retail drugstores, wholesale drug distributors, service stations, oil distributors, industrial chemical suppliers, and repackagers of chemicals.

MINERAL SPIRITS (Petroleum Naphtha): Clear liquid, from the petroleum distillation process. CAUTION: Flammable. Available from service stations, oil distributors, industrial chemical suppliers, and repackagers of chemicals.

MONOCALCIUM PHOSPHATE: See Superphosphate.

MONTAN WAX (Lignite Wax): White, hard, earth wax, soluble in benzene. Melting point 90° C. Made by extraction of lignite from coal. Available from retail drugstores, wholesale drug distributors, industrial chemical suppliers, and repackagers of chemicals.

NAPHTHA (V.M.P.): White, highly volatile liquid, made in the petroleum distillation process. CAUTION: Highly flammable. Available from oil distributors, industrial chemical suppliers, and repackagers of chemicals.

NAPHTHALENE (Tar Camphor) $C_{10}H_8$: White crystalline flakes, soluble in benzene, absolute alcohol, and ether. Made by boiling coal tar oil and then crystallizing. Available from retail drugstores, wholesale drug distributors, industrial chemical suppliers, and repackagers of chemicals.

NEATSFOOT OIL: A pale yellow oil, soluble in alcohol and kerosene. Made by boiling in water the shinbones and feet, without hoofs, of cattle. The oil and fat are then separated. Available from feed and grain suppliers, oil distributors, and repackagers of chemicals.

NEROLI OIL (Orange Flower Oil): Amber-color oil, soluble in equal parts of alcohol. Made by the distillation of citrus flowers. Available from retail drugstores, wholesale drug distributors, industrial chemical suppliers, and repackagers of chemicals.

NICOTINE SULFATE ($C_{10}H_{14}N_2$)2.H_2SO_4: White crystals, soluble in water or alcohol. Made by the action of sulfuric acid on the alkaloid. CAUTION: Toxic. Available from industrial chemical suppliers.

OIL OF MUSTARD: The extract of mustard seed (Brassica spp.) through distillation. Available from wholesale drug distributors, chemical repackagers and importers of essential oils.

OIL OF VANILLA: An extract obtained from the fruit or bean of a species of orchid native to Central America and Mexico. Available from repackagers of chemicals.

OIL SOAP: An emulsion used by machine shops for lubricating metal cutting tools. Available from retail paint and hardware stores, wholesale hardware stores, industrial chemical suppliers and paint suppliers.

OLEIC ACID (Red Oil) $CH_3(CH_2)_7CH:CH(CH_2)_7COOH$: Yellow to red oily liquid, soluble in alcohol and organic solvents. Derived from animal tallow or vegetable oils. Available from retail drugstores, wholesale drug distributors, industrial chemical suppliers, and repackagers of chemicals.

OLIVE OIL: Pale yellow to greenish liquid, nondrying. Only slightly soluble in alcohol. Soluble in ether, chloroform, or carbon disulfide. Oil is cold-pressed from the olive fruit and then refined. Available from retail and wholesale groceries and repackagers of chemicals.

OLEORESIN CAPSICUM: Oleoresin is a semisolid mixture of the resin and the essential oil of the plant from which it is derived, *Capsicum* (cayenne pepper, African pepper, red pepper). Dried fruit of *Capsicum frutescens, Capsicum annuum*, or the Louisiana sport pepper. Available from retail drugstores, wholesale drug distributors, industrial chemical suppliers, and repackagers of chemicals.

ORANGE OIL (Citrus Seed Oil): Orange-color oil, expressed from orange seeds. Bitter taste is removed by refining. Available from retail drugstores, wholesale drug distributors, industrial chemical suppliers, and repackagers of chemicals.

ORTHODICHLOROBENZENE $C_6H_4CL_2$: Colorless, heavy liquid, miscible with most organic solvents. Made by chlorinating monochlorobenzene. CAUTION: Moderately toxic by ingestion, but highly irritating to skin and eyes. Available from retail drugstores, wholesale drug distributors, industrial chemical suppliers, and repackagers of chemicals.

OXALIC ACID $HOOCCOOH.2H_2O$: Transparent, colorless crystals, formed in nature by the oxidation of proteins in plants such as wood sorrel, rhubarb, and spinach. CAUTION: Toxic. Available from retail drugstores, wholesale drug distributors, industrial chemical suppliers, and repackagers of chemicals.

OXYQUINOLINE SULFATE $(C_9H_7NO)_2H_2SO_4$: Pale yellow powder, soluble in water. CAUTION: Moderately toxic in concentrated form. Available from retail drugstores, wholesale drug distributors, industrial chemical suppliers, and repackagers of chemicals.

PALM OIL (Palm Butter): Yellow-brown soft, solid material, soluble in alcohol. Derived from nuts and fruit of the palm tree native to West Africa. Available from retail drugstores, wholesale drug distributors, industrial chemical suppliers, and repackagers of chemicals.

PARACHLOROMETACRESOL $CH_3C_6H_4OH$: Colorless to yellow liquid, soluble in alcohol. Derived from coal tar. Available from industrial chemical suppliers.

PARADICHLOROBENZENE $C_6H_4Cl_2$: White volatile crystals, soluble in alcohol, benzene, and ether. Made by chlorination of monchlorobenzene. Moderately toxic by ingestion. Irritant to eyes. Also known as moth crystals. Available from industrial chemical suppliers.

PANCREATIN: Cream-color powder, soluble in water. Extracted from the pancreas of cattle and hogs. Available from wholesale drug distributors, industrial chemical suppliers, and repackagers of chemicals.

PARAFFIN OIL: An oil pressed from paraffin distillate. For characteristics, see Paraffin Wax. Available from service stations, oil distributors, industrial chemical suppliers, and repackagers of chemicals.

PARAFFIN WAX: White, waxy blocks, soluble in benzene, warm alcohol, turpentine, and olive oil. Made by distilling crude petroleum oil. Available from retail and wholesale groceries, oil distributors, industrial chemical suppliers, and repackagers of chemicals.

PEANUT OIL (Groundnut Oil): Yellow oil, soluble in petroleum, ether, carbon disulfide, and chloroform. Can be saponified by alkali hydroxides to form a soap. Available from retail and wholesale groceries, oil distributors, and repackagers of chemicals.

PENTACHLOROPHENOL C_6Cl_5OH: White powder or crystals, soluble in alcohol. Made by chlorinating phenol. CAUTION: Highly toxic by ingestion, inhalation, and absorption through the skin. Available from industrial chemical suppliers, chemical repackagers and wholesale drugs.

PEPPERMINT OIL: Clear, oily liquid, soluble in alcohol. Made by distilling the leaves of the peppermint plant. Available from retail drugstores, wholesale drug distributors, and repackagers of chemicals.

PETROLEUM DISTILLATE: Colorless, volatile liquid, miscible with most organic solvents and oils. Made by distillation from petroleum. Available from oil distributors, industrial chemical suppliers, and repackagers of chemicals.

PETROLATUM (Mineral Wax, Petroleum Jelly, Mineral Jelly): Colorless to amber translucent oil mass, soluble in benzene, ether, chloroform, and oil. Melting point 60° C. Made by the distillation of still residues from steam distillation of paraffin-based petroleum. Available from retail drugstores, wholesale drug distributors, oil distributors, industrial chemical suppliers, and repackagers of chemicals.

PHENOL: See Carbolic Acid.

PINE OIL: Colorless to amber oily liquid. Miscible with alcohol. Made by steam distillation of pine wood. Available from retail drugstores, wholesale drug distributors, industrial chemical suppliers, and repackagers of chemicals.

PINE TAR: Resinous product derived from the sap of pine trees usually obtained as a by-product of turpentine distillation. Available from retail drugstores, wholesale drug distributors, industrial chemical suppliers, and repackagers of chemicals.

PORTLAND CEMENT: White to gray powder composed of lime, alumina, silica, and iron oxide. Available from building supply dealers.

POTASH (Potassium Carbonate, Pearl Ash) K_2CO_3: White deliquescent translucent powder, soluble in water. CAUTION: Toxic if taken internally. Available from feed and grain suppliers, oil distributors, and repackagers of chemicals.

POTASSIUM CARBONATE: See Potash.

POTASSIUM CHLORIDE (Murate of potash) KCl: White solid, soluble in water with a melting point of 790°C: important as potash fertilizer. Available from repackagers of chemicals and industrial chemical suppliers.

POTASSIUM HYDROXIDE: See Caustic Potash.

POTASSIUM NITRATE (Niter, Saltpeter) KNO_3: Transparent or white crystals or powder, soluble in water. CAUTION: Dangerous fire and explosion risk when subjected to shock or heating. Oxidizing agent. Handle carefully. Available from retail drugstores, wholesale drug distributors, industrial chemical suppliers, and repackagers of chemicals.

POTASSIUM OLEATE $C_{17}H_{33}COOK$: Gray to tan paste, soluble in water and alcohol. Available from retail drugstores, wholesale drug distributors, industrial chemical suppliers, and repackagers of chemicals.

POTASSIUM PERMANGANATE: Permanganate and potash, purple solid soluble formed by oxidation of ossified potassium manganate solution. CAUTION: Caustic, do not allow contact with skin. Available from retail drugstores, wholesale drug distributors, industrial chemical suppliers, and repackagers of chemicals.

POTASSIUM PERSULFATE $K_2S_2O_8$: White crystals, soluble in water. Made by electrolysis of a saturated solution of potassium sulfate. CAUTION: Moderately toxic. Available from retail drugstores, wholesale drug distributors, industrial chemical suppliers, and repackagers of chemicals.

POTASSIUM SULFATE (Potash Alum) $AL_2(SO_4)_3 K_2SO_4.24H_2O$: White crystals, soluble in water. Derived from the mineral alunite. Available from retail drugstores, wholesale drug distributors, industrial chemical suppliers, and repackagers of chemicals.

POWDERED GRAPHITE: An allotropic form of carbon produced artificially by heating coal or coke; ground into powdered form. Available from retail and wholesale paint and hardware distributors, industrial chemical suppliers, and ceramic shops.

POWDERED SKIM MILK: Powdered milk is prepared by dehydrating whole milk. Available from retail drugstores, wholesale drug distributors, and retail and wholesale groceries.

POWDERED SULFUR: Nonmetalic mined mineral. See flowers of sulfur.

PRECIPITATED CHALK: Limestone washed out with water. Available from building supply dealers, chemical repackagers, ceramic shops and retail and wholesale paint and hardware stores.

PROPYLENE GLYCOL $CH_3CHOHCH_2OH$: Colorless viscous liquid, odorless and tasteless. Miscible with water and alcohol. Made by hydration of propylene oxide. Available from wholesale drug distributors, oil distributors, industrial chemical suppliers, and repackagers of chemicals.

PROPYLENE GLYCOL MONSTEARATE: A mixture of propylene glycol mono- and diesters of stearic acid. Soluble in alcohol, chloroform, and other chlorinated hydrocarbons. White flakes or beads, bland odor and taste. CAUTION: Combustible. Available from industrial chemical suppliers, and wholesale drug distributors.

PUMICE POWDER: A gently abrasive fine powder milled from porous rock found in nature. Available from retail drugstores, wholesale drug distributors, industrial chemical suppliers, and repackagers of chemicals.

PYRETHRUM EXTRACT: A powder obtained from milling ground pyrethrum flowers. Usually mixed with kerosene or other solvents. CAUTION: Moderately toxic if taken internally. Available from retail drugstores, wholesale drug distributors, feed and grain suppliers, industrial chemical suppliers, and repackagers of chemicals.

QUARTZ ScO_2: Oxide of the non-metallic element silicon, commonest of metals. Available from building supply dealers, chemical repackagers, ceramic shops and retail and wholesale paint and hardware stores.

QUASSIA (Bitterwood, Bitter Ash): White to bright yellow chips. Takes on a tannish color when ground. Obtained from the bark of *Picrasma* or *Quassia* trees. Available from retail drugstores, wholesale drug distributors, and repackagers of chemicals.

RAPE SEED OIL (Colza Oil, Rape Oil): Pale yellow viscous liquid, soluble in ether, chloroform, and carbon disulfide. Made by expression or solvent extraction of rape seeds. Available from oil distributors, industrial chemical suppliers, and repackagers of chemicals.

RAW LINSEED OIL: Untreated oil from the flaxseed presses which is filtered through duck and flannel cloths in a plate and frame press. Yellow-brown or amber in color. Develops heat spontaneously. Available from retail and wholesale paint and hardware stores, industrial chemical suppliers, building supply dealers and repackagers of chemicals.

REFINED TALLOW: An animal fat extracted from the solid fat or "suet" of cattle, sheep, or horses by dry or wet rendering. Available from retail and wholesale groceries.

ROSEMARY OIL: Clear to slightly yellowish oil, soluble in alcohol, ether, and glacial acetic acid. Made by steam-distilling the flowers of *Rosmarinus officinalis*. Available from retail drugstores, wholesale drug distributors, industrial chemical suppliers, and repackagers of chemicals.

ROSIN: Translucent amber chips, soluble in alcohol, ether, glacial acetic acid, and oil. Derived by steam distillation of the sap of pine trees. Available from retail drugstores, wholesale drug distributors, industrial chemical suppliers, and repackagers of chemicals.

ROTTENSTONE (Tripoli): White abrasive powder. Crushed and milled from rock. Available from retail and wholesale paint and hardware stores, industrial chemical suppliers, and repackagers of chemicals.

SAL SODA (Washing Soda, Sodium Carbonate Decahydrate) $Na_2CO_3 \cdot 10H_2O$: White crystals soluble in water, insoluble in alcohol. Pure form of sodium carbonate (soda ash). CAUTION: Moderately toxic; irritant to mucous membranes. Wear rubber gloves when handling. Available from retail and wholesale paint and hardware stores, industrial chemical suppliers, and repackagers of chemicals.

SALICYLIC ACID (Orthohydroxy Benzoic Acid) $C_6H_4(OH)(COOH)$: White powder, soluble in alcohol, oil of turpentine, and ether. Made by treating a hot solution of sodium phenolate with carbon dioxide. Available from retail drugstores, wholesale drug distributors, industrial chemical suppliers, and repackagers of chemicals.

SAND (Silicon Dioxide) SIO_2: White crystals or powder, soluble only in hydrofluoric acid or molten alkali. Found widely in its natural state. Can also be made from a soluble silicate (waterglass) by acidifying, washing, and igniting. Available from building supply dealers.

SASSAFRAS LEAVES: The leaves of the plant *Sassafras albidum*. Available from retail drugstores, wholesale drug distributors, industrial chemical suppliers, and repackagers of chemicals.

SESAME OIL (Benne Oil, Teel Oil): Bland yellow liquid, soluble in chloroform, ether, and carbon disulfide. Extracted from the plant *Sesamum indicum*, found in China, Japan, and South America. Available from retail drugstores, wholesale drug distributors, industrial chemical suppliers, and repackagers of chemicals.

SHELLAC (Garmet Lac, Gum Lac, Stick Lac): A natural resin secreted by the laccifer and deposited on trees in India. Soluble in alcohol. Available from retail and wholesale paint and hardware stores, industrial chemical suppliers, and repackagers of chemicals.

SILICA GEL: Hard white lumps, crystals, or powder. Regenerative adsorbent, having a vast internal porosity in relation to its outside surface. Made by the reaction of sulfuric acid and sodium silicate. Available from repackagers of chemicals and refrigeration supply dealers.

SILICONE WATER EMULSION: A milky, slippery liquid that can be further diluted with water to any desired concentration. Made by the mixture of silicone oil, emulsifier, and water. Available from repackagers of chemicals or a foundry supplier.

SILVER NITRATE $AgNO_3$: Transparent crystals, soluble in cold water or hot alcohol. Made by dissolving silver in dilute nitric acid and evaporating. CAUTION: Highly toxic. Strong irritant; handle with care. Available from retail drugstores, wholesale drug distributors, or repackagers of chemicals.

SLACKED LIME: Reaction of limestone (chiefly calcium carbonate) with water in an exothermic process. Available from industrial chemical suppliers, wholesale paint and hardware stores, building supply dealers and ceramic shops.

SODA ASH (Sodium Carbonate) NA_2CO_3: Grayish white powder, soluble in water. Mined in areas such as Great Salt Lake, or can be made by the Solvay ammonia soda process. Available from retail and wholesale paint and hardware stores, and repackagers of chemicals.

SODIUM ALGINATE ($C_6H_7O_6Na$): Colorless to light yellow solid. May be in granular or powdered form. Forms a thick collodial solution with water. Made by extraction from brown seaweed (kelp). Available from retail drugstores, wholesale drug distributors, industrial chemical suppliers, or repackagers of chemicals.

SODIUM ALUMINATE $Na_2Al_2O_4$: White powder, soluble in water. Made by heating bauxite with sodium carbonate and extracting the sodium aluminate with water. Available from retail drugstores, wholesale drug distributors, industrial chemical suppliers, and repackagers of chemicals.

SODIUM BARATE (borax) $Na_2B_4O_7.10H_2O$: Formed by reaction of sodium carbonate with boric acid, calcium carbonate or purification of the mineral. Available from retail drugstores, supermarkets, and wholesale paint and hardware stores.

SODIUM BICARBONATE (Baking Soda) $NaHCO_3$: White powder, soluble in water. Made by treating a saturated solution of soda ash with carbon dioxide. Available from retail and wholesale paint and hardware stores and from retail groceries.

SODIUM BISULFATE (Niter Cake) $NaHSO_4$: Colorless crystals soluble in water. A by-product in the manufacture of hydrochloric and nitric acids. CAUTION: Toxic in solution. Irritant to eyes and skin. Available from retail drugstores, wholesale drug distributors, industrial chemical suppliers, and repackagers of chemicals.

SODIUM CARBONATE: See Soda Ash.

SODIUM CHLORATE $NaClO_3$: Colorless, odorless crystals, soluble in water and alcohol. Made by heating and electrolyzing a concentrated solution of sodium chloride. Available from oil distributors, industrial chemical suppliers, and repackagers of chemicals.

SODIUM CHLORIDE (Salt) $NaCl$: White crystals soluble in water and glycerol. Made by the evaporation of salt brine. Available from retail and wholesale groceries and feed and grain suppliers.

SODIUM CITRATE $C_6H_5O_7Na_3.2H_2O$: White crystals or powder, soluble in water. Made by treating a sodium sulfate solution with calcium citrate. CAUTION: Combustible.

SODIUM DIDELARO Striazinetrione: White powder. Decomposes in water.

SODIUM DODECYLBENZENE SULFATE $C_{12}H_{25}C_6H_4SO_3Na$: White to light yellow flakes or powder. Biodegradable. Available from retail drugstores, wholesale drug distributors, industrial chemical suppliers, and repackagers of chemicals.

SODIUM HYDROXIDE: See Caustic Soda.

SODIUM HYPOCHLORITE $NaOCl$: Pale greenish solution soluble in cold water. Made by the addition of chlorine to a cold dilute solution of sodium hydroxide. CAUTION: Toxic by ingestion and inhalation. Irritant to skin and eyes. Available from retail drugstores, wholesale drug distributors, industrial chemical suppliers, and repackagers of chemicals.

SODIUM LAURYL SULFATE $NaC_{12}H_{25}SO_4$: White or light yellow crystals, soluble in water. Acts as a wetting agent. Available from retail drugstores, wholesale drug distributors, industrial chemical suppliers, and repackagers of chemicals.

SODIUM METAPHOSPHATE $NaPO_3$: White powder, soluble in water. Available from retail drugstores, wholesale drug distributors, industrial chemical suppliers, and repackagers of chemicals.

SODIUM METASILICATE Na_2SiO_3: A crystalline silicate. White granules, soluble in water. Available from retail drugstores, wholesale drug distributors, industrial chemical suppliers, and repackagers of chemicals.

SODIUM NITRATE (Saltpeter) $NaNO_3$: Colorless, transparent crystals, soluble in water and glycerol. Made from nitric acid and sodium carbonate. CAUTION: Moderately toxic and flammable. Explodes when subjected to physical shock or temperatures of 1000° F. Available from retail drugstores, wholesale drug distributors, industrial chemical suppliers, and repackagers of chemicals.

SODIUM PENTACHLOROPHENATE C_6Cl_5ONa: White to tan powder, soluble in water and acetone. CAUTION: Toxic by ingestion and inhalation. Irritant to eyes and skin. Use gloves when handling. Available from retail drugstores, wholesale drug distributors, industrial chemical suppliers, and repackagers of chemicals.

SODIUM PERBORATE $NaBo_2.3H_2O$: White, odorless powder crystals. Decomposes in water to release oxygen. Made by electrolysis of a solution of borax and soda ash. Available from retail drugstores, wholesale drug distributors, industrial chemical suppliers, and repackagers of chemicals.

SODIUM PHOSPHATE NA_2HPO_4: White powder, soluble in water and alcohol. Made by precipitating calcium carbonate from a solution of dicalcium phosphate with soda ash. Available from retail drugstores, wholesale drug distributors, industrial chemical suppliers, and repackagers of chemicals.

SODIUM PYROPHOSPHATE $Na_4P_2O_7$: Transparent crystals or white powder. Soluble in water. Made by fusing sodium phosphate. Available from retail drugstores, wholesale drug distributors, industrial chemical suppliers, and repackagers of chemicals.

SODIUM SESQUICARBONATE (Sesqui) $NaCO_3.NAHCO_3 2H_2O$: White needle-shaped crystals, soluble in water. Made by crystallation from a solution of sodium carbonate and sodium bicarbonate. Available from retail drugstores, wholesale distributors, industrial chemical suppliers, and repackagers of chemicals.

SODIUM SILICATE (Waterglass) $Na_2O.3.75SiO_2$: Clear viscous liquid, soluble in water. Made by the fusion of sand and soda ash. CAUTION: May be irritating and caustic to skin and mucous membranes. Available from retail drugstores, wholesale drug distributors, industrial chemical suppliers, and repackagers of chemicals.

SODIUM SULFATE (Salt Cake) Na_2SO_4: White crystals or powder, soluble in water. A by-product of hydrochloric acid production from salt and sulfuric acid.

Available from retail drugstores, wholesale distributors, industrial chemical suppliers, and repackagers of chemicals.

SODIUM SULFITE Na_2SO_3: White crystals or powder, soluble in water. Made by reacting sulfur dioxide with soda ash and water. Available from retail drugstores, wholesale drug distributors, industrial chemical suppliers, and repackagers of chemicals.

SODIUM THIOSULFATE (Hypo) $Na_2S_2O_3.5H_2O$: White crystals or powder, soluble in water and oil of turpentine. Made by heating a solution of sodium sulfite with powdered sulfur. Available from retail drugstores, wholesale distributors, industrial chemical suppliers, and repackagers of chemicals.

SODIUM TRIPOLYPHOSPHATE (Tripoly) $Na_2S_2O_3.5H_2O$: White powder, soluble in water. Made by calcination of sodium orthophosphate mixture from sodium carbonate and phosphoric acid. Available from retail drugstores, wholesale drug distributors, industrial chemical suppliers, and repackagers of chemicals.

SOFT SOAP: A liquid soap made with potassium hydroxide and a vegetable oil (except coconut and palm kernel oil). Available from retail drugstores, wholesale distributors, industrial chemical suppliers, and repackagers of chemicals.

SOYBEAN OIL (Chinese Bean Oil, Soy Oil): Pale yellow drying oil, soluble in alcohol, ether, chloroform, or carbon disulfide. Made by expression and solvent extraction of crushed soybeans. Available from retail and wholesale groceries, feed and grain suppliers, oil distributors, and repackagers of chemicals.

SPERMACETI: White, semitransparent, waxlike solid, soluble in ether, chloroform, carbon disulfide, and hot alcohol. Extracted from the head of the sperm whale. Available from retail drugstores, wholesale drug distributors, and repackagers of chemicals.

STANNIC OXIDE SnO_2: White powder, soluble in sulfuric acid and hydrochloric acid. Found in nature in the mineral cassiterite. Available from retail drugstores, wholesale drug distributors, oil distributors, industrial chemical suppliers, and repackagers of chemicals.

STEARAMIDE $CH_3(CH_2)_{16}CONH_2$: Colorless flakes, soluble in alcohol. Available from retail drugstores, wholesale drug distributors, industrial chemical suppliers, and repackagers of chemicals.

STEARIC ACID $CH_3(CH_2)_{16}CO_2H$: Waxlike solid, soluble in alcohol, ether, chloroform, or carbon disulfide. Made by hydrogenation of oleic acid. Available from drugstores, wholesale drug distributors, oil distributors, industrial chemical suppliers, and repackagers of chemicals.

STODDARD SOLVENT: Water-white liquid solvent. CAUTION: Mildly flammable. Available from retail drugstores, oil distributors, repackagers of chemicals, and dry cleaners' suppliers.

SUGAR: A carbohydrate product of photosynthesis comprising one, two, or more saccharose groups. Available from retail and wholesale groceries.

SULFITE LIQUOR: A waste liquor from the sulfite paper-making process. Synthetic vanilla (vanalin) is made from this material. Available from oil distributors and repackagers of chemicals.

SULFITE LYE: See Sodium Hydroxide.

SULFONATED CASTOR OIL: A vegetable oil that has been treated with sulfuric acid and neutralized with a small amount of caustic soda. The oil is then emulsifiable with water. Available from retail drugstores, wholesale drug distributors, oil distributors, industrial chemical suppliers, and repackagers of chemicals.

SULFUR, WETTABLE: Pure sulfur exists in two stable crystalline forms. Alpha-sulfur: rhombic, octahedral, yellow crystals stable at room temperature. Beta-sulfur: monoclinic, prismatic, pale yellow crystals. Both forms are insoluble in water; slightly soluble in alcohol and ether; soluble in carbon disulfide, carbon tetrachloride, and benzene. Available from industrial chemical suppliers and repackagers of chemicals.

SULFURIC ACID (Hydrogen Sulfate, Oil of Vitroil, Battery Acid) H_2SO_4: Strongly corrosive, dense, oily liquid, colorless to dark brown, depending on purity. Miscible with water in all proportions. Dissolves most metals; concentrated acid oxidizes, dehydrates, or sulfonates most organic compounds. CAUTION: Use caution in mixing with water. Always add the acid to water, never the reverse. Highly toxic and strong irritant to tissue. Wear gloves and use in well-ventilated area. Available from retail drugstores, wholesale drug distributors, industrial chemical suppliers, and repackagers of chemicals.

SULFONATED CASTOR OIL: A vegetable oil that has been treated with sulfuric acid and neutralized with a small amount of caustic soda. The oil is then emulsifiable with water. Available from retail drugstores, wholesale drug distributors, oil distributors, industrial chemical suppliers, and repackagers of chemicals.

SUPERPHOSPHATE (Acid Phosphate) $CaH_4(PO_4)_2 \cdot H_2O$: Water-soluble powder, made by the action of sulfuric acid on insoluble rock. Available from feed and grain suppliers and repackagers of chemicals.

TALC (Talcum, Mineral Graphite, Steatite) $Mg_3Si_4O_{10}(OH)_2$: A mined mineral (magnesium silicate), white-gray pearly color with a greasy feel. Available from retail drugstores, wholesale drug distributors, oil distributors, and repackagers of chemicals.

TANNIC ACID (Gallotainic Acid, Tannin) $C_{76}H_{52}O_{46}$: Light yellow crystals, soluble in water, alcohol, and benzene. Made by extraction of nutgalls and tree bark, with water and alcohol. Available from retail drugstores, wholesale drug distributors, oil distributors, industrial chemical suppliers, and repackagers of chemicals.

TAPIOCA FLOUR (Amylum) $(C_6H_{10}O_5)$: White amorphous, tasteless powder; irregular lumps or fine powder. Insoluble in cold water, alcohol, and ether; forms a gel with hot water. Derived from cassava (tapioca). Available from retail and wholesale groceries.

TETRASODIUM PYROPHOSPHATE (TSP): Colorless crystals or white powder, soluble in water. Made by fusing sodium phosphate. Available from oil distributors, industrial chemical suppliers, and repackagers of chemicals.

TARTARIC ACID (CREAM OF TARTAR) $HOOC(CHOH)_2COOH$: White crystalline powder, soluble in water and alcohol. Made from maleic anhydride and hydrogen peroxide. Available from retail drugstores, wholesale drug distributors, industrial chemical suppliers, and repackagers of chemicals.

TIN, POWDERED (STANNUM): White ductile solid. Metallic element of atomic number 50; group IVA of the periodic system. Available from industrial chemical suppliers and repackagers of chemicals.

TINCTURE OF ARNICA: Medication derived from a plant, usually in tinctures of various strengths. CAUTION: Toxic by ingestion. Available from retail drugstores, wholesale drug distributors, industrial chemical suppliers, and repackagers of chemicals.

TINCTURE OF BENZOIN $C_6H_5CHOHCOC_6H_5$: Clear to pale yellow liquid, having a slight camphor odor. The crystals from which the tincture is made are derived from the condensation of benzaldehyde in a cyanide solution. Available from retail drugstores, wholesale drug distributors, industrial chemical suppliers, and repackagers of chemicals.

TINCTURE OF IODINE: Water and alcohol mixture of potassium iodine used medically as an antiseptic. CAUTION: Concentration of iodine may increase as alcohol evaporates. Keep tightly capped, avoiding older containers that have been previously opened; iodine can cause severe burns in its concentrated forms. Available from retail drugstores, wholesale drug distributors, industrial chemical suppliers, and repackagers of chemicals.

TINCTURE OF RHUBARB: Dried root and stalks of rhubarb are treated with alcohol to form a tincture (about 10% solution). Available from retail drugstores, wholesale drug distributors, industrial chemical suppliers, and repackagers of chemicals.

TITANIUM DIOXIDE (Titanium White, Titania) TiO_2: White powder, miscible with water, alcohol, or oil. Made by treating ilmenite with sulfuric acid. Available from retail drugstores, wholesale drug distributors, oil distributors, and repackagers of chemicals.

TITANIUM TRICHLORIDE $TiCl_3$: Dark violet crystals, soluble in alcohol. Flammable in the presence of oxidizing materials. CAUTION: Skin irritant. Available from industrial chemical suppliers and chemical repackagers.

TOLUENE (Toluol, Methylbenzene) $CH_3C_6H_5$: White liquid, soluble in alcohol, benzene, and ether. Made by fractional distillation of coal tar oil. CAUTION: Flammable. Available from retail and wholesale paint and hardware stores, industrial chemical suppliers, and repackagers of chemicals.

TRAGACANTH (Tragacanth Gum): White flakes or yellow powder. Soluble in alkaline solution. Available from retail drugstores, wholesale drug distributors, industrial chemical suppliers, and repackagers of chemicals.

TRICALCIUM PHOSPHATE CA_2HPO_4: A white powder or solid, soluble in water and alcohol CAUTION: May be somewhat irritating to the skin. Available from industrial chemical suppliers.

TRICHLOROETHYLENE (Tri, Trichlor) $CHCl:CCl_2$: Colorless, heavy liquid; slightly soluble in water, miscible with organic solvents. Made from tetrachloroethane by treatment with alkali in the presence of water. CAUTION: Vapors are toxic. Use with adequate ventilation. Available from retail drugstores, wholesale drug distributors, retail and wholesale paint and hardware stores, industrial chemical suppliers, and repackagers of chemicals.

TRIETHANOLAMINE (TEA, TRI) $(HOCH_2CH_2)_3N$: Colorless viscous hygroscopic liquid, miscible with water and alcohol. Made by the reaction of ethylene oxide and ammonia. CAUTION: May be somewhat irritating to skin and mucous membranes. Available from retail drugstores, wholesale drug distributors, industrial chemical suppliers, and repackagers of chemicals.

TRISODIUM PHOSPHATE (Sodium Phosphate Dibasic) NA_2HPO_4: Colorless crystals or white powder, soluble in water and alcohol. Made by precipitating calcium carbonate from a solution of dicalcium phosphate with soda ash. CAUTION: Skin irritant, use rubber gloves. Moderately toxic by ingestion. Available from retail and wholesale paint and hardware stores, industrial chemical suppliers, and repackagers of chemicals.

TURPENTINE $C_{10}H_{16}$: Colorless, clear, oily liquid. Made by steam distillation of turpentine gum. CAUTION: Toxic if taken internally. Flammable. Handle with care. Available from retail and wholesale paint and hardware stores and industrial chemical suppliers.

ULTRAMARINE BLUE: Blue lumps, soluble in oil. Made by heating a mixture of sulfur, clay, alkali, and reducer at high temperatures. Available from retail drugstores, wholesale drug distributors, industrial chemical suppliers, and repackagers of chemicals.

UREA (Carbamide) $CO(NH_2)_2$: White crystals or powder, soluble in water, alcohol, and benzene. Derivation: liquid ammonia and liquid carbon dioxide react under pressure and elevated temperatures to form ammonium carbonate, which decomposes at lower temperatures to form urea. Available from retail drugstores, wholesale drug distributors, industrial chemical suppliers, and repackagers of chemicals.

UREA PHOSPHATE: See Urea.

VENICE TURPENTINE (Larch Gum): Yellow to greenish resin, soluble in most organic solvents. Distilled from *Larix europaca*. Available from retail and wholesale paint and hardware stores and industrial chemical suppliers.

VERMICULITE: Crystalline-type structure with high porosity. Insoluble, except in hot acids. Used as filler in concrete and for thermal insulation. Available from building supply dealers.

VINEGAR (Dilute Acetic Acid): Brown liquid, dilutable with water. Made by fermentation of fruits and grains. May be distilled to remove brown color, after which it is known as white vinegar. Available from retail and wholesale groceries, industrial chemical suppliers, and repackagers of chemicals.

WASHING SODA: See borax.

WHITE BEESWAX: Wax from the honeycomb of frames in the beehive. White color is obtained by bleaching the natural yellow wax. Soluble in chloroform, ether, and oils. Melting point 62°-65° C. Available from retail drugstores, wholesale drug distributors, industrial chemical suppliers, and repackagers of chemicals.

WITCH HAZEL: A clear white, astringent liquid, soluble in water and alcohol. Available from retail drugstores, wholesale drug distributors, industrial chemical suppliers, and repackagers of chemicals.

WOOD TAR (Pine Tar): Viscous, sticky, brown-to-black syrup, soluble in alcohol and acetone. Made by the destructive distillation of pine wood. Available from retail drugstores, wholesale drug distributors, industrial chemical suppliers, and repackagers of chemicals.

XYLOL (Xylene) $C_6H_4(CH_3)_2$: Clear liquid, soluble in alcohol and ether. Made by fractional distillation of petroleum, coal tar, or coal gas. CAUTION: Flammable. Available from oil distributors, industrial chemical suppliers, and repackagers of chemicals.

YELLOW BEESWAX: See White Beeswax. Note: Yellow and white beeswax have the same properties except color. Therefore, where color is not important (as in floor wax, for example), the yellow wax is more economical. Available from retail drugstores, wholesale drug distributors, industrial chemical suppliers, and repackagers of chemicals.

YELLOW DEXTRIN: Yellow or white powder, soluble in water. Made by heating dry starch or by treating starch with dilute acid. Available from wholesale paint and hardware stores and industrial chemical suppliers.

ZINC Zn: Shining white metal of atomic number 30. Soluble in acids and alkalies. Mined in British Columbia, Utah, Colorado, Idaho, Peru, and Australia. Available from ceramic shops and repackagers of chemicals.

ZINC ACETATE $Zn(C_2H_3O_2)_2.2H_2O$: White crystalline plates, soluble in water and alcohol. Made by the action of acetic acid on zinc oxide. Available from ceramic shops and repackagers of chemicals.

ZINC BROMIDE $ZnBr_2$: White crystalline powder, soluble in water, alcohol, and ether. Made by the interaction of solutions of barium bromide and zinc sulphate, and then crystallized. Available from retail drugstores, wholesale drug distributors, industrial chemical suppliers, and repackagers of chemicals.

ZINC CHLORIDE $ZnCl_2$: White crystals or crystalline powder, soluble in water, alcohol, and glycerin. Made by the action of hydrochloric acid on zinc. CAUTION: Toxic. Available from retail drugstores, wholesale drug distributors, industrial chemical suppliers, and repackagers of chemicals.

ZINC OXIDE (Chinese White, Zinc White) ZnO: Coarse white to gray powder, soluble in acids and alkalies. Made by oxidation of vaporized pure zinc. CAUTION: Poisonous if taken internally. Available from retail drugstores, wholesale drug distributors, industrial chemical suppliers, and repackagers of chemicals.

ZINC STEARATE $Zn(C_{18}H_{35}O_2)_2$: Soluble in common solvents and hot acids. Made by the action of sulfuric acid on zinc. Available from retail drugstores, wholesale drug distributors, industrial chemical suppliers, and repackagers of chemicals.

ZINC SULFATE (White Vitriol) $ZnSO_4.7H_2O$: Colorless crystalline powder, soluble in water or glycerol. Made by the action of sulfuric acid on zinc oxide. Available from retail drugstores, wholesale drug distributors, industrial chemical suppliers, and repackagers of chemicals.